Foundations of Chemistry
Applying POGIL Principles
Fifth Edition

David M. Hanson

Stony Brook University - SUNY

John Goodwin

Coastal Carolina University

Maria Phillips

New College of Florida

FOUNDATIONS OF CHEMISTRY
Applying POGIL Principles

Fifth Edition

by   David M. Hanson (Stony Brook University, SUNY)

John Goodwin (Coastal Carolina University)

Maria Phillips (New College of Florida)

Pacific Crest

P.O. Box 370 Hampton, NH 03843

603-601-2246 www.pcrest.com

ISBN: 978-1-60263-516-6

The companion website to this text is available at:
www.foc5.com

Acknowledgments

- David Hanson would like to thank Dan Apple, founder and president of Pacific Crest and acknowledge him as the motivating force and inspiration that led to *Foundations of Chemistry* and the associated process-oriented guided-inquiry learning movement, which now is known as POGIL. His insights on activity design and classroom facilitation are much appreciated by many. He would also like to thank again all those who contributed to and were acknowledged in previous editions and especially the coauthors and those mentioned below who produced this fifth edition.

- John Goodwin would like to acknowledge his general chemistry students and department colleagues at Coastal Carolina University for their feedback regarding POGIL activities and my use of them for nearly twenty years over multiple editions. *"I appreciate the support I have received from students and colleagues at Coastal Carolina University in my use of POGIL and active learning in courses that are typically taught in a very traditional manner."*

- Maria Phillips is credited with for revising and creating new Exercises and Problems for the *Student edition* and drafting answer keys and facilitation notes for the *Instructor's Edition*. Maria also provided invaluable proofreading feedback. *"I would like to acknowledge all of the wonderful support I have received from my family throughout my life, I would not be where I am today without them. I would also like to thank the many mentors I have had the privilege of learning from; you have shaped me into the teacher and scientist that I am."*

- The quality of the fifth edition was enhanced significantly through the skills of Denna Hintze (Pacific Crest, editing and production) and Breanna Apple (Pacific Crest, copy editing).

- Support from the National Science Foundation made it possible to develop, test, and revise activities for general and physical chemistry; share them with others; help others move from lecturing to more student-centered teaching strategies; and work toward enhancing process oriented guided inquiry activities with computer and web technology. The following grants supported this and related projects: DUE-9752570, DUE-9950612, DUE-0127650, DUE-0127291, DUE-0231120, and DUE-0341485.

- The electrochemical cell graphic in Activity 18.1 is based on Figure 2 in 17.3 *Standard Reduction Potentials* by Rice University and used according to Creative Commons Attribution 4.0 International License.

TABLE OF CONTENTS

TO THE INSTRUCTOR AND STUDENT

Process Oriented Guided Inquiry Learning (POGIL) is both a philosophy of and strategy for teaching and learning. It is a philosophy because its practice is based on specific ideas about the nature of the learning process as well as the expected outcomes of learning. It is a strategy because it provides a specific structure for teaching that is consistent with the way people learn, thus leading to the desired outcomes.

The goal of POGIL is to engage students in the learning process, helping them master the material through development of conceptual understanding as they gradually build upon a foundation of important new ideas, terminology, and mathematical expressions regarding physical and chemical relationships. Students who carefully prepare at this **most basic** level of learning will then be able to apply their foundational learning in situations that are more and more challenging. As understanding of the principles and relationships grows, students are able to solve new and unfamiliar problems. In repeated modeling of this approach to learning it is intended that students will gain an appreciation for the necessity of organizing their learning in progressive stages.

Important skill areas for success in chemistry courses, college, and beyond include: information processing, critical and analytical thinking, problem solving as emphasized in the structure of the written activities, but just as importantly oral and written communication, teamwork, and metacognition (reflection on learning, self-assessment, and self-management). These skills are emphasized in the structure of the students' work before group activities, engagement with their classmates in the classroom, and finally their extension of that work into mastery of the material in follow-up assignments, assessments, and exams. Ultimately, POGIL activities utilize a learning cycle design that consists of exploration, concept formation, and subsequent application.

Each activity begins with an *Orientation* that sets the stage for learning. The importance of the activity is described in a *Why* statement (*Why* is short for "Why should I care about this?" or "Why should I bother learning this?"). *Learning Objectives* and *Success Criteria* are identified along with prerequisite activities. The *Learning Objectives* describe what the student is expected to learn through completing the activity. The *Success Criteria* specify the measurable outcomes of this learning; they describe how the learner should be able to demonstrate that he or she has successfully learned the activity content. It is relatively quick and easy to write quiz or test questions simply by using the *Success Criteria* for each activity. The *Prerequisites* identified in an activity are the previous activities that students should have successfully completed as preparation for the current activity. Students should not merely gloss over these, but take the time to review the *Learning Objectives* from the prerequisite activities, ensuring they are ready to tackle the new learning challenges.

Students then explore a *Model* or *Information* section through responding to *Key Questions*. (A *Model* is any representation of what is to be learned.) Students are encouraged to critically examine Models in anticipation of answering *Key Questions*. Doing so reinforces the value of reading for understanding. *Key Questions* are the kinds of questions students should ask themselves as they approach new information so that they can discover its meaning and significance. *Information* sections are provided at the beginning and throughout an activity to contextualize and reinforce the significance of the *Models*.

The time needed to acquire the new knowledge in the *Models* and *Information* sections may necessitate students working through them as pre-class assignments. It is reasonable to expect that students review prerequisite activities and carefully read and reflect on what is new (terminology, concepts, mathematical symbols, graphs, etc.) before working on the *Exercises*, which serve reinforce the new knowledge and build learner confidence. The *Problems* and *Got-It!* Sections require learners to synthesize ideas, generalize their learning by transferring it to new contexts, and demonstrate their problem-solving skills. While many of the *Problems* may seem like familiar "examples" that an instructor would demonstrate in a traditional lecture-format class, the focus here is on students solving the *Problems* independently of an instructor's lead. This makes them more challenging and ultimately more memorable and engaging for students, thereby making them more effective.

While students *can* complete the activities working individually, the activities are most effective when students work through them in learning teams with an instructor acting as a coach, guide, and facilitator.

There are many to help instructors teach in this type of student-centered environment. They can be found at the Pacific Crest (www.pcrest.com) and POGIL (www.pogil.org) web sites. Both Pacific Crest and the NSF-supported POGIL Project sponsor workshops for faculty to introduce them to process oriented guided inquiry learning, assist them in developing facilitation skills, and guide them to materials for use in their courses.

TO THE STUDENT

Changes in society, technology, and the world economy are occurring at increasingly faster rates. As a college graduate, you will need to be a quick learner, critical thinker, and successful problem solver to succeed in this rapidly changing environment. You will also need to be computer literate and demonstrate a high degree of skill in the areas of communication, teamwork, management, and assessment. This book is intended for use in courses where faculty are responding to these pressures and needs by making changes in the way they teach and using curricula that will actively engage you in learning, helping you to develop these essential skills.

While this book may or may not be one of the printed books you are required to buy for your college classes, it is probably the only one you're asked to actually write in. While it provides you with information to read, terms, symbols, and concepts to memorize, and models to interpret at a basic level of understanding, its real value is in how increases your understanding *beyond* this level to one where you can respond successfully to exercises and problems rather than simply repeating procedures and sequences of calculations that can be memorized without any understanding.

Just as importantly, the activities in this book model effective learning process skills. This book and the activities within it have been very carefully designed with the goal of helping you learn how to process information, analyze situations through inquiry (asking yourself key questions), construct your understanding of chemistry, and develop the problem solving skills that will help you be successful in this course, in college, and beyond.

You will learn the most and have the most fun if you work on these activities with other students. Discussions among members of your learning team will produce different perspectives regarding the concepts and their use in solving problems. These discussions will help you to identify and correct misconceptions you may have and strengthen and deepen your understanding of chemistry. But *individual* preparation is required before you can participate in an intelligent discussion. If you come to class without having reviewed prerequisites and completed any pre-class reading assignments, you will be at an immediate disadvantage and will not be able to contribute to or gain from any discussion of the material. As a result, you will probably disengage from the discussion and waste your time as well as that of your classmates.

Use your textbook to resolve disagreements, find answers to questions that arise, and see examples of solutions to problems. But it is through your understanding of the concepts and how to *use* them that you will be able to successfully answer exam questions and solve real-world problems. When you are working in a learning team, you should have two objectives: to understand the material yourself and to ensure that every other member of the team understands the material as well. Explaining ideas and helping others learn are among the best ways for you to deepen your own understanding, giving you valuable insight (and skills) that will further ensure your success on exams, as well as when you're faced with real-world problem that must be solved.

We have found that this approach works for most students: they do better on exams, develop a deeper understanding of chemistry, recognize that they have become stronger learners, and have more fun along the way.

WHY?

Units identify the scale that is used in making a measurement and are essential for the measurement to be meaningful. For example, if someone tells you that a person is "60 tall," you do not know whether they are referring to a child or an adult until you know the units. The units could be inches or centimeters (but probably not feet or meters). In your study of chemistry and its applications, you need to be familiar with the basic units that are used for mass, length, time, temperature, electrical current, and amount of substance. Unit prefixes make it quicker and easier to write very large or very small numbers, e.g., 5,600,000 g = 5.6 Mg or 0.000001 s = 1 μs. Other units are derived from these basic units. For example, units of volume are derived from units of length, and units of energy are derived from units of mass, length, and time. You also need to be able to convert from one set of units to another because different countries, disciplines, and even sub-disciplines of chemistry often use different units for the same quantity. For example, in the United States, speed limits are given in miles per hour; in many other countries speed limits are given in kilometers per hour.

LEARNING OBJECTIVES

- Identify the units used to measure physical quantities and dimensions
- Become familiar with the prefixes used for larger and smaller quantities based on powers of ten
- Master the use of unit conversion based on defined equivalencies in solving problems

SUCCESS CRITERIA

- Associate units with physical quantities or dimensions
- Replace prefixes by multiplying by appropriate numerical factors
- Identify, set up, compute, and validate unit conversions based on defined equivalencies

PREREQUISITE

- Exponential or scientific notation

MODEL 1   INTERNATIONAL SYSTEM OF UNITS (SI UNITS)

Table 1

Physical Quantity	Name of Unit	Abbreviation
mass	kilogram	kg
length	meter	m
time	second	s
temperature	Kelvin	K
electrical current	ampere	A
amount of substance	mole	mol

(When you see this dotted line, it means that the model is continued on the following page.)

Table 2

Prefix	Abbreviation	Meaning
	T	10^{12}
	G	10^{9}
	M	10^{6}
	k	10^{3}
	c	10^{-2}
	m	10^{-3}
	μ	10^{-6}
	n	10^{-9}
	p	10^{-12}
	f	10^{-15}

Examples

Mass: A quarter-pound hamburger has a mass of 0.11 kg. 0.11 kg = 110 g

Length: A tall basketball player (7 feet) has a height of 2.15 m. 2.15 m = 215 cm = 2150 mm

Time: There are 3300 s in a 55 minute chemistry lecture. 3300 s = 3.300 ks

Temperature: Water freezes at 273 K.

1×10^{3} g = 1000 g = 1 kg

1 cm = 0.01 m = 1×10^{-2} m

10 cm × 10 cm × 10 cm = 1000 cm^3 = 1 L (Note: L = liter)

1 ps = 1×10^{-12} s

KEY QUESTIONS

1. What are the 6 basic units and their abbreviations listed in **Model 1** (Table 1)? (You need to have these units and their abbreviations memorized. Practicing writing them out without looking at the table will help you remember.)

2. What are the 10 prefixes listed in **Model 1** (Table 2), along with their abbreviations and meanings? (Work on memorizing these prefixes, their abbreviations, and meanings.)

Chapter 1: Chemistry is a Quantitative Science

3. How is the unit of volume (liter) derived from the unit of length (cm) in one of the Examples below **Model 1**? What is an equivalent expression for volume based on combined units of length?

4. My snowfall amount, reported in New York: 22. My friend's snowfall amount, as reported in Germany (in the same period of time): 50. But I got more snow that he did. How is that possible?

EXERCISES

1. Write the number of minutes in a day (1,440 min) in exponential notation.

2. Write the abbreviation for the units used when the number of seconds in a day is expressed as 86.4

3. The average life of a molecule of poitronium hydride is 0.5 ns. Write this time using exponential notation with units of s.

4. Actor Robert DeNiro famously gained a great deal of weight to play the role of Jake La Motta in Raging Bull. Write the mass he gained, approximately 27,000 grams, in units of kg.

5. According to Wikipedia, the mass of the Batmobile used in the 1960s TV series is about 2.1 Mg. Write this mass using exponential notation with units of g.

6. The diameter of a ribosome is about 20 nm. Write this length using exponential notation with units of m.

7. The diameter of Mars is about 6,800 km. Write its diameter using units of mm.

8. How many fs are there in 1s?

| Model 2 | A GENERAL CHEMISTRY PROBLEM-SOLVING STRATEGY APPLIED TO CONVERTING UNITS |

Following this six-step problem-solving strategy will help you solve more complicated problems. It is illustrated here for the case of unit conversion.

In a unit conversion problem, you need an *equality statement* that tells you how the two units are related, e.g., 1 quarter = 25 pennies or 2.2 lb = 1.0 kg.

From the equality statement, you can produce *conversion factors* that are used to convert one unit into the other, e.g., 1 quarter/25 pennies or 1.0 kg/2.2 lb.

Example Unit Conversion Problem: You work in a pharmacy and need to convert the mass of a medicine from ounces to drams. There are 5.00 ounces of the medicine; what is the mass in drams?

Strategy	Example
Step 1: Identify and record what is known or given.	**Known:** have 5.00 ounces of medicine
Step 2: Identify and record what is unknown and needs to be found.	**Unknown:** mass of medicine in drams
Step 3: Identify and record the concepts that connect what needs to be found to what is known.	**Connections:** look up an equality statement: 1 ounce = 15.75 drams conversion factors: 1 ounce/15.75 drams = 1 15.75 drams/1 ounce = 1
Step 4: Set up the solution using the connections.	**Setup:** apply the appropriate conversion factor 5.00 ounces × (15.75 drams/1 ounce)
Step 5: Perform the calculations to obtain the result.	**Result:** 78.8 drams (note that the units cancel)
Step 6: Check or validate your answer.	**Validation:** units correct, value is reasonable (~5 × 16)

KEY QUESTIONS

5. What units need to be converted in **Model 2**?

6. How many conversion factors result from a single equality statement in the model? Explain.

Chapter 1: Chemistry is a Quantitative Science

7. When the conversion factor was applied in Step 4 to obtain the result in Step 5, what happened to the ounces unit?

8. In general, how can you identify whether you have used the correct unit conversion factor?

9. Is multiplying by a unit conversion factor the same as or different from multiplying by 1? Explain.

EXERCISES

9. If Robert De Niro had a mass of 215 lbs when he starred in Raging Bull, what was his mass in kg? (1.00 kg = 2.20 lbs)

10. Did I get more snow than my friend in Germany? I got 24.3 inches; he got 64.8 cm. (1 in = 2.54 cm).

11. Complete the following table.

Equality Statement	Conversion	Conversion Factor	Result
2.54 cm = 1.00 in	100 in to cm	$\dfrac{(2.54\ cm)}{(1.00\ in)}$	
1.00 L = 1.06 qt	5.5 qt to L	$\dfrac{(1.00\ L)}{(1.06\ qt)}$	
9.0 tibbles = 42.1 squips	75 squips to tibbles	$\dfrac{(9.0\ tibbles)}{(42.1\ squips)}$	

PROBLEMS

1. Your lab bench is 42 inches by 130 inches. How many square meters of acid-resistant film will you need to cover the bench? (Use 1.0 m = 39 inches.)

WHY?

Unit analysis (or *dimensional analysis*) is a procedure that produces the units associated with answers to mathematical calculations. It facilitates problem solving, validates the solutions, and sometimes involves unit conversions. Engineers, health professionals, biologists, and other scientists often rely on unit analysis in their work.

LEARNING OBJECTIVES

- Recognize equivalencies based on defined relationships between units, measured physical constants, and situations given in a problem
- Master the use of dimensional analysis in making unit conversions, solving problems, and validating answers

SUCCESS CRITERION

- Set up, compute, and validate the solution (units and magnitude) of computational problems

PREREQUISITE

01-1 *Units of Measurement*

MODEL 1 — A GENERAL CHEMISTRY PROBLEM-SOLVING STRATEGY APPLIED TO DIMENSIONAL ANALYSIS

In the country of Apocrib, force is measured in units of nocums and area is minum squared. Find the pressure (both the units and the magnitude) when a force of 12 nocums is exerted on a window pane with an area of 24 minum^2. Pressure is defined as force divided by area, $P = F/A$.

Methodology	Example
Step 1: Identify and record what is known or given.	**Known:** force = 12 nocums area = 24 minum^2
Step 2: Identify and record what is unknown and needs to be found.	**Unknown:** pressure
Step 3: Identify and record the concepts that connect what needs to be found to what is known.	**Connections:** pressure = force / area $P = F/A$
Step 4: Set up the solution using the connections.	**Setup:** $P = F/A$ $P = 12$ nocums / 24 minum^2
Step 5: Perform the calculations to obtain the result.	**Result:** perform division to both the numbers and the units, obtain 0.50 nocums / minum^2
Step 6: Check or validate your answer.	**Validation:** Units are correct, value is correct (12 / 24 = 0.50)

With contributions by Vicky Minderhout, Seattle University

KEY QUESTIONS

1. What is the purpose of each step in the problem solving strategy in **Model 1**?

2. In a calculation, how can you obtain the units to associate with the numerical answer?

3. How can you use unit analysis to identify whether you have performed numerical algebraic operations correctly?

4. What meaning is lost when units are omitted? What are the general implications of not using units?

A General Chemistry Problem-Solving Strategy Applied to Dimensional Analysis: Combined Units in Measured Constants and Unique Experimental Situations

MODEL 2

When expressing measured physical quantities, combined units are often required. For example, the speed of light in a vacuum has a measured value of 2.9979×10^8 m/sec and the number of particles in 1 mole is 6.022×10^{23} particles/mole. Densities of pure substances are another type of constant that combine mass (in g) and volume (in cm^3). Defined relationships between units within a common dimension are used to create conversion factors.

For example, in the dimension of length, 1 meter is equivalent to 100 centimeters, giving the conversion factor $m = \dfrac{cm}{100}$.

This also works for combined units such as density, where a conversion factor can be used to convert from a known quantity to an unknown quantity.

Example: If gold is known to have a density of 19.32 g/cm^3, what would be the mass of 10.00 cm^3 of gold?

Methodology	Example
Step 1: Identify and record what is known or given.	**Known:** volume = 10.00 cm^3 density = 19.32 g/cm^3
Step 2: Identify and record what is unknown and needs to be found.	**Unknown:** mass (in g)
Step 3: Identify and record the concepts that connect what needs to be found to what is known.	**Connections:** g/cm^3 (the units given for density in this case), is mass divided by volume. We can use the known density and volume to solve for mass: $$density = \frac{mass}{volume}$$
Step 4: Set up the solution using the connections.	**Setup:** $$19.32 \text{ g / cm}^3 = \frac{mass}{10.00 \text{ cm}^3}$$
Step 5: Do the mathematics to obtain the result.	**Result:** mass = 10.00 cm^3 × (19.32 g/cm^3) = 193.2 g
Step 6: Check or validate your answer.	**Validation:** Units are correct, value is correct (193.2 / 10 = 19.32)

1. Density is mass divided by volume. If mass is measured in units of vectas and volume is measured in units of tarts, determine the units of density in this unit system.

2. An 1859 book, *The Corner Cupboard*, lists prices and common measurements in use at that time. It states that a *firkin* of beer costs $1.39.

    a. What conversion factors can be created from the information given?

    b. What conversion factors can be created from the following statements?
        *   1 *firkin* = exactly 9 gal
        *   1 *tierce* = exactly 42 gal

    c. Are all of these conversion factors always true, like those created from physical constants? Explain your answer.

    d. How much would you pay in dollars for a *tierce* of beer?

    e. You may have successfully solved part d in a series of steps using single conversion factors. You could have also solved this problem using a series of conversion factors multiplied together in a chain of terms. If you did not do so in part d, repeat the set-up of your solution as **a series of conversion factors**. This method saves time and is generally clearer in showing the overall cancellation of units. (Unless you've got a lot of time on your hands and nothing better to do, consider using this approach!)

3. Chemists have found that $PV = nRT$ for gases where P = pressure in atmospheres, V = volume in liters, n = amount of gas in moles, R = a constant, and T = temperature in Kelvin. Identify the units only of the constant R. Note that units on both sides of an equal sign must match.

4. *Pressure* is defined as force per unit area ($P = F/A$), and *force* is defined as mass times acceleration ($F = ma$). In the SI system, pressure is measured in units of Pascals (Pa). If the following SI units are used: mass = kg, acceleration = m/s^2, and area = m^2, write the dimensional equality statement relating 1 Pa to the SI units of kg, m, and s.

5. Explain how unit analysis validates your answers to the following.

   a. The mass of a diamond is measured in carats. What is the volume of a 1.30 carat diamond if its density is 3.50 g/cm^3? (1 carat = 0.200 g) Use a chain of conversion factors in setting up your solution.

   b. In 2009 Usain Bolt set the world record for the 100 m race at 9.58 seconds. What was Usain's average speed in mi/hr? (1610 meters = 1 mile)

INFORMATION

You will find the following information helpful in solving the problem below.

Sodium fluoride, NaF, is 45.0% fluoride by mass. *Percent by mass* is defined as the number of parts out of 100 parts of the total mixture. So in this case it means *45.0 g fluoride per 100.0 g sodium fluoride* which implies the conversion factor of 45.0 g fluoride/100.0 g sodium fluoride.

The word *per* means "divided by" or "for every", so the statement, "45.0 g fluoride per 100.0 g sodium fluoride" could be translated as, "45.0 g fluoride for every 100.0 g sodium fluoride."

1 ppm means "one part per million" by mass, e.g., 1 g of fluoride per one million grams of water. 1 gallon = 3.79 L

1 year = 365 days	1 ton = 2000 lb	1 lb = 453.6 g	density of water = 1.0 g/mL

PROBLEM

You are the public health officer in the water treatment facility for a city of 95,000 people. A concentration of 1.0 ppm of fluoride in the drinking water is sufficient for the purpose of helping to prevent tooth decay. The compound normally chosen for fluoridation is the same as is found in some toothpastes: sodium fluoride, NaF. Calculate how many kilograms of sodium fluoride you will need to purchase in order to fluoridate the city's water supply for one year, based upon your estimate that the average yearly consumption of water is 110 gallons per person.

Why?

The number of digits or figures reported for a numerical quantity conveys the quality or precision of the measurement or numerical result of a calculation involving measurements to the reader. In the laboratory, the number of significant figures is determined by the instrument or glassware used to make the measurement and includes the first estimated digit as the last significant figure in the reported value. In this course and in others, and in your career, you will be expected to use a meaningful number of digits in reporting your numerical results. Your textbook clearly outlines rules for reporting significant figures and the advantages of exponential notation in avoiding ambiguous numbers of significant figures. Be especially careful in applying the rules related to leading zeros after a decimal point and trailing zeros before a decimal point.

Learning Objectives

- Appreciate the difference between accuracy and precision
- Understand the relationship between precision and the number of significant figures in a number

Success Criteria

- Identify the accuracy and precision of a numerical value
- Report computed values to the correct number of significant figures

Prerequisites

01-1 *Units of Measurement*

01-2 *Unit Analysis*

Information

Accuracy is the degree of conformity to a standard or true value.

Precision is the smallest repeatable (consistent) digit in a series of measurements of the same quantity.

Significant figures are the repeatable digits and the first uncertain digit in a measurement or calculation.

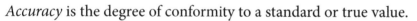

Model 1   Accuracy, Precision, and Significant Figures

Item	Values	Significant Figures
Bureau of Standards Time	9:15:13.004	8
Jerry's watch	9:15	3
Jennifer's iPhone	9:17:52	5
average mass of gold coin	23.3 g	3
height of an index card	0.0770 m	3

1. In measuring time, which value in the model represents the standard value?

2. Is Jerry's or Jennifer's measurement of time more accurate? Explain.

3. Is Jerry's or Jennifer's measurement of time more precise? Explain.

4. How is precision represented when reporting a measurement?

5. Why do we say that the height of the index card is reported to three significant figures?

6. In a laboratory experiment, what are two ways to improve

    a. the precision of a measurement?

    b. the accuracy of a measurement?

EXERCISES

1. Specify the number of significant figures in a) through f) below. Identify any cases where there may be ambiguity in the number of significant figures and how you could clarify the precision of the measurement.

    a. 1001.1 cm

    b. 0.0275 cm

    c. 1000 cm

d. 1.000×10^2 cm

e. 1000. cm   (Note: including the decimal point is a convention to show that the number is 1000 and not 999 or 1001)

f. 0.0043000 cm

2. Express the measurement 500 cm so it is clear it has only two significant figures.

PROBLEMS

1. The mass of a gold coin was measured three times and each measurement was made to five digits. The mass values were 23.319 g, 23.341 g, and 23.296 g. The average mass was reported as 23.3 g. The actual mass of the coin is 25.5631 g.

   a. Are these measurements precise? Explain your answer.

   b. Are these measurements accurate? Explain your answer.

   c. Why is the average mass of the gold coin reported to only three significant figures?

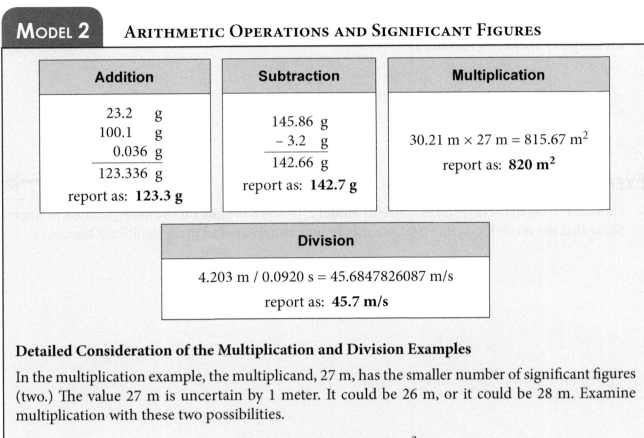

Detailed Consideration of the Multiplication and Division Examples

In the multiplication example, the multiplicand, 27 m, has the smaller number of significant figures (two.) The value 27 m is uncertain by 1 meter. It could be 26 m, or it could be 28 m. Examine multiplication with these two possibilities.

$$30.21 \text{ m} \times 26 \text{ m} = 785.46 \text{ m}^2$$

$$30.21 \text{ m} \times 28 \text{ m} = 845.88 \text{ m}^2$$

The three results differ in the tens place, so the product of the two numbers is reported to only two significant digits, 820 m², more clearly expressed as 8.2×10^2 m². Notice that the product has the same number of significant figures as the *least* certain multiplicand.

In Exercise 3, which follows, you will be asked to analyze the division example in a way similar to the example in **Model 2**, to show that the quotient should be reported with three significant figures.

KEY QUESTIONS

7. When you add or subtract numbers as shown in **Model 2**, how can you identify the first uncertain digit in the result?

8. When you multiply or divide numbers, what is the relationship between the number of significant digits in the result and the number of significant figures in the numbers you are multiplying or dividing as illustrated in **Model 2**?

EXERCISES

3. Consider in detail the division example in **Model 2**, just as was done for the multiplication example. Show that the result for 4.203 / 0.0920 s should only be reported to three significant figures.

4. Report the total mass of three people weighing 56 kg, 59.1 kg, and 72.89 kg. Explain the rationale for the number of significant figures in your answer.

5. Calculate the density (mass/volume) of a coin with a mass of 7.9 g and a volume of 0.981 cm^3. Explain why you should report the result to two significant figures and not to one or three.

02-1 Atoms, Isotopes, and Ions

WHY?

Atoms are the fundamental building blocks of all substances. To begin to understand the properties of atoms and how they combine to form molecules and ionic compounds, you must be familiar with how three sub-atomic particles — protons, neutrons, and electrons — are arranged in atoms, isotopes, and ions.

LEARNING OBJECTIVES

- Be able to characterize three sub-atomic particles: protons, neutrons and electrons
- Understand the composition and structure of atoms, isotopes, and ions
- Understand how atomic symbols and names identify the number of sub-atomic particles composing an atom, isotope, or ion

SUCCESS CRITERIA

- Use atomic symbols to represent different isotopes and ions
- Given one or more of the following items, determine the others: name, atomic symbol, atomic number, mass number, neutron number, and electron number
- Calculate the percent of the atomic mass that is located in the nucleus of an atom
- Compare the size of an atom to the size of the atomic nucleus

PREREQUISITES

Calculation of percent

01-1 *Units of Measurement*

01-2 *Unit Analysis*

INFORMATION

Matter, which is anything that has mass and occupies space, is composed of substances and mixtures of substances.

A *substance*, or more explicitly, a *pure substance*, is a variety of matter that has uniform and constant composition down to the molecular level. For example, pure water is a substance.

Mixtures are composed of two or more substances. For example, salt water is a mixture, even though it is uniform, because the amount of salt in the water (the composition) can vary.

Sub-atomic particles important for chemistry are the *proton*, *neutron*, and *electron*. Protons and neutrons are located in the *nucleus* of the atom and have about the same mass. Protons have a positive charge of +1, but neutrons have zero charge. Electrons have a very small mass by comparison, a negative charge of –1, and are located outside of the nucleus.

An *element* is a substance that cannot be decomposed into two or more other substances by chemical or physical means. In nuclear reactions, however, one element can be converted into one or more other elements. Only 118 different elements are known to exist (as of 2016) and these are listed on the periodic table of the elements with their atomic numbers (Z): the number of protons in each atom of a given element. The number of protons in the nucleus is equal to the total positive charge of an element and defines what that element is. Hydrogen, H, with one proton (atomic number 1), and carbon, C, with six protons (atomic number 6), are both examples of elements.

An *atom* is the smallest amount of an element that can exist either alone or in combination with other atoms, forming *molecules* that are held together by chemical bonds between the atoms. Bonds are made up of new arrangements of the electrons from the atoms. For example, one carbon atom, C, and two oxygen atoms, O, can bond together to create the carbon dioxide molecule, CO_2.

Isotopes are atoms of a given element that have the same number of protons but different numbers of neutrons, N. Carbon has two abundant isotopes, carbon-12 and carbon-13, but small amounts of carbon-14 are also found in nature. All of these carbon isotopes contain six protons in the nucleus, since the atomic number of carbon is six. They have different *mass numbers, A*, which are the sum of the number of protons and neutrons in an isotope. Mass numbers are different from average atomic masses listed on the periodic table. These will be discussed later.

 The mass of an isotope CANNOT be determined by adding up the masses of the individual protons, neutrons, and electrons that are combined, due to subtleties arising from nuclear chemistry.

An *ion* is an atom or molecule with a positive or negative charge that arises from an unequal number of negatively charged electrons and positively charged protons. Atoms in their elemental state have equal numbers of protons and electrons. Monatomic ions are formed by adding or removing electrons. An atom of sodium has 11 protons and 11 electrons in its elemental state.

A *cation* is an ion with a positive charge resulting from having fewer electrons than protons. An *anion* is an ion with a negative charge resulting from having more electrons than protons.

MODEL 1 — NOTATION AND REPRESENTATIONS OF SODIUM

The diagrams which follow show representations of sodium. Note that the diameter of an atom is about 10,000 times larger than the diameter of the atomic nucleus.

Table 1

Particle	Mass (amu)*	Charge
proton	1.0073	+1
neutron	1.0087	0
electron	0.0005	−1

* *Atomic mass unit* (amu) is a unit of mass equal to 1.66054×10^{-27} kg.

ATOMIC SYMBOL NOTATION

Figure 1

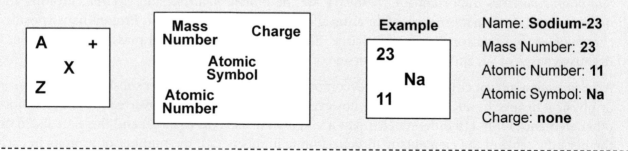

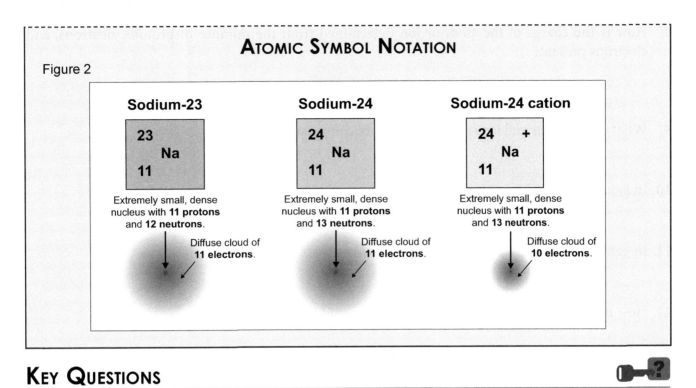

Figure 2

ATOMIC SYMBOL NOTATION

KEY QUESTIONS

1. What are the three particles that comprise a sodium atom?

2. Which particles contribute most of the mass to the atom, and where are these particles located?

3. Which particles contribute most to the volume or size of the atom, and where are these particles located?

4. What information is provided by the atomic number, Z, which is the subscript in the atomic symbol?

5. What information is provided by the mass number, A, which is the superscript in the atomic symbol?

6. What notation is used in the atomic symbol to indicate the charge of an atom or ion?

7. Given the definition of mass number and the information in Table 1 regarding the masses of protons, neutrons, and electrons, why is the mass number approximately, but not exactly equal to, the mass of an atom in amu?

8. How is the charge of the atom or ion determined from the number of protons, neutrons, and electrons present?

9. What do all atoms and ions of sodium have in common?

10. In general, what feature of an atom identifies it as a particular element?

11. In general, how do isotopes of the same element differ?

12. How many isotopes of any particular element could there be?

EXERCISES

1. Add any missing information to the following table. The first row is completed for you as an example.

Name	Symbol	Z	A	Number of Neutrons	Number of Electrons
boron-10	$^{10}_{5}B^{+}$	5	10	5	4
	$^{40}_{20}Ca$				18
oxygen-16					8
	U			146	92
		9	19		9
		17		18	18
			39	20	19

2. Show how to calculate the mass of a proton, neutron, and electron in kilograms using the data in Table 1 and the equality statement **1 amu = 1.66054 × 10^{-27} kg**

PROBLEMS

1. The mass of a carbon-12 atom is 12 amu.

   a. What sub-atomic particles are contained in this atom?

   b. What percent of the mass of this atom is located inside the nucleus?

   c. What percent of the mass is therefore located outside the nucleus?

   d. Why is the value you calculated for the nuclear mass so close to 100%?

   e. Why is the sum of all the individual particle masses NOT equal to 12 amu?

12	
	C
6	

2. The radius of a Cl nucleus is 4.0 fm, and the radius of a Cl atom is 100 pm. If the nucleus of the Cl atom were the size of a dime, which is 17 mm in diameter, determine whether the atom would be approximately a) the size of a quarter, b) the size of a car, c) the size of a football stadium, or d) the size of the earth. Explain how you made your decision.

WHY?

The mass of an atom in amu is approximately equal to its mass number, but a **mass spectrometer** can be used to determine the exact masses of atoms. This powerful instrument reveals the presence of isotopes and measures the naturally occurring proportion of each isotope of an element. With this information, the average mass of all the isotopes of an element can be determined. It is this average mass that is listed on the periodic table. Knowing the average mass of an element in atomic mass units allows us to use the concept of the *mole*, which chemists use on a daily basis.

Since the relative masses of the elements are constant for a given number of atoms, we scale up *atomic masses in amu* to masses of *one mole of atoms in grams* using the same average mass values from the periodic table. That is, one atom of carbon-12 has a mass of exactly 12 amu, by definition. The mass of one mole of carbon-12 atoms has a mass of exactly 12 g, by definition. The number of atoms required to make this conversion from one atom to one mole is Avogadro's number, which is equal to 6.022×10^{23} atoms/mole. Knowing how many atoms or molecules are in a sample is essential to understanding mass relationships in chemical formulas and reactions. This area of chemistry is called *stoichiometry* and will be studied in detail later.

LEARNING OBJECTIVE

- Understand the basic idea of mass spectrometry, and how the mass of an atom can be determined

SUCCESS CRITERIA

- Successfully interpret a mass spectrum
- Successfully construct a mass spectrum from appropriate data
- Convert between mass in amu and mass in grams
- Determine the number of atoms in a given mass of an element

PREREQUISITES

01-1 *Units of Measurement*

01-2 *Unit Analysis*

01-3 *Significant Figures in Data*

02-1 *Atoms, Isotopes, and Ions*

INFORMATION

In a mass spectrometer, an electrical discharge knocks electrons off the atoms or molecules, giving them a net positive charge. These ions then are accelerated in an electric field, and some property related to their mass is measured. In some mass spectrometers, the property is the trajectory of the ions in a magnetic field. In other mass spectrometers, the property is the time taken by the ions to travel the distance from the point of ionization to a detector. An example of a mass spectrum is shown in **Model 1**.

In a mass spectrum the number of ion counts over some period of time is plotted on the *y*-axis and the atomic mass is plotted on the *x*-axis. The number of ion counts for different isotopes is proportional to the abundance of each isotope in the sample. Since atomic masses are so small, a special *mass* unit is used. This unit is called the *atomic mass unit* and is abbreviated *amu*. The *atomic mass unit* is defined so the mass of the carbon-12 isotope is exactly 12 amu. The masses of all other atoms then are measured relative to the mass of carbon-12. The factor for converting amu to grams is 1.66054×10^{-24} g/amu.

MODEL 1   DATA ON THE MASSES OF ATOMS

Mass Spectrum of Boron

Figure 1

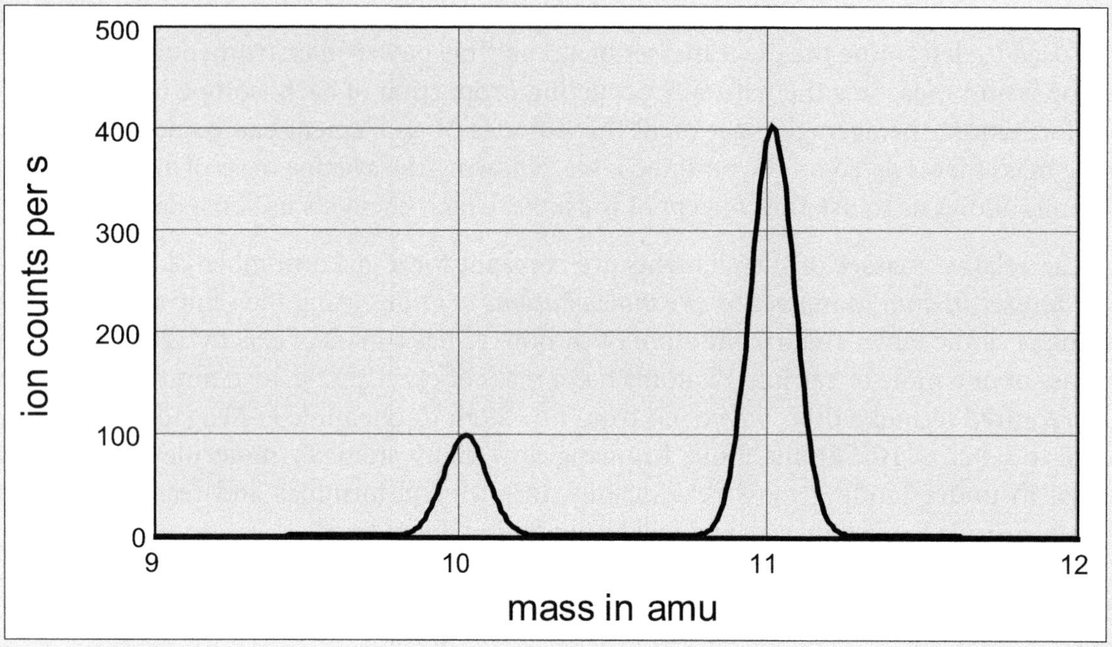

Isotopic Data

Table 1

Atomic Isotope	Symbol	Mass (amu)*	Natural Abundance on Earth (%)
Hydrogen	$^{1}_{1}H$	1.0078	99.985
Deuterium	$^{2}_{1}H$	2.0140	0.015
Helium	$^{4}_{2}He$	4.00260	100.0
Boron-10	$^{10}_{5}B$	10.0129	19.78
Boron-11	$^{11}_{5}B$	11.0093	80.22
Carbon-12	$^{12}_{6}C$	exactly 12 (by definition)	98.89
Carbon-13	$^{13}_{6}C$	13.0034	1.11
Chlorine-35	$^{35}_{17}Cl$	34.9689	75.53
Chlorine-37	$^{37}_{17}Cl$	36.9481	24.47

KEY QUESTIONS

1. What mass unit is used to report the mass of atoms in the model?

2. According to the *Information* section, how is this mass unit defined?

3. Why are there two peaks in the mass spectrum of boron?

4. Why is one peak in the mass spectrum for boron higher than the other?

5. Are the positions of the peaks on the x-axis and the ratio of peak heights in the boron mass spectrum consistent with the data in *Table 1: Isotopic Data*? Explain.

6. The mass number and atomic mass both have a value of 12 for carbon-12. Do isotopes of all other elements also have the same value for the mass number and atomic mass? Explain.

7. Why is it more convenient to use atomic mass units (amu) rather than grams or kilograms (g or kg) when reporting masses of individual atoms?

8. If an element has several isotopes, what single value for the mass could be used to characterize atoms of that element? Justify your answer.

EXERCISES

1. Calculate the ratio of boron-11 to boron-10 found on Earth using the information given in Table 1.

2. Describe how you can determine the ratio of boron-11 to boron-10 found on Earth from information provided by the mass spectrum of boron.

3. From the information in Table 1, determine the number of boron-11 atoms that you would expect to find in a natural sample of 10,000 boron atoms.

4. Sketch what you might expect the mass spectrum of a natural sample of atomic chlorine to look like.

5. If you could pick one carbon atom from a natural sample, what would its mass in grams most likely be? Explain.

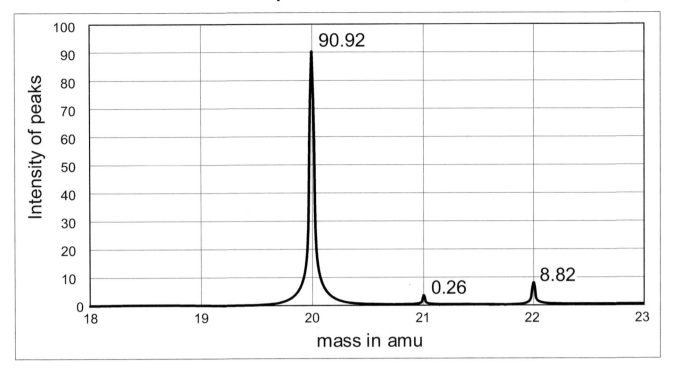

Mass Spectrum of Neon   Figure 2

Intensity of peaks (y-axis): 0, 10, 20, 30, 40, 50, 60, 70, 80, 90, 100

90.92

0.26

8.82

mass in amu (x-axis): 18, 19, 20, 21, 22, 23

6. From the mass spectrum for neon shown in Figure 2, determine

   a. The number of nickel isotopes present in the sample

   b. The mass numbers of the nickel isotopes

   c. The number of neutrons in each isotope

   d. The relative abundance of each isotope

   e. What single value for the mass could be used to characterize atoms of neon?

ACTIVITY
02-3 The Periodic Table of the Elements

WHY?

Substances that contain only atoms with the same number of protons are called *elements*. The Periodic Table lists all the known elements in order of increasing atomic numbers, Z, which is the characteristic number of protons for each element. The values of Z increase from left to right and from top to bottom across the periodic table. The columns of the Periodic Table contain elements with similar chemical and physical properties, and as we will see much later, the arrangement of electrons in atoms is the underlying basis for these properties. The Periodic Table is a useful tool for both students and professionals that can be used to identify the properties of the elements and understand the properties of molecules and ionic compounds.

LEARNING OBJECTIVES

- Become familiar with the organization of the Periodic Table
- Appreciate both the diversity and commonalities in the chemical and physical properties of the elements

SUCCESS CRITERIA

- Identify groups and periods in the Periodic Table
- Use the Periodic Table to provide information about the elements

PREREQUISITES

02-1 *Atoms, Isotopes, and Ions*

02-2 *Mass Spectrometry and Masses of Atoms*

INFORMATION

Dmitri Mendeleev (1834 – 1907), a Russian scientist, constructed the first Periodic Table by listing the elements in horizontal rows in order of increasing atomic mass. He started new rows whenever necessary to place elements with similar properties in the same vertical column. Mendeleev found that the correlations in properties between some elements in the columns were not perfect. These observations led him to predict the existence of undiscovered elements to fill gaps in his version of the Periodic Table with known elements and to wonder how the table might be better organized. Later H. G. J. Moseley used x-ray spectra to refine the ordering and show that atomic numbers rather than atomic masses should be used to order the elements.

In the Periodic Table, elements with similar properties occur in vertical columns called *groups*. Two numbering conventions are used to label the groups. The older convention numbers the groups using Roman numerals I through VIII followed by a letter A or B; the other convention numbers each column 1 through 18. The A groups are known as the *main group elements*. The B groups are called the *transition elements*. The group numbers IA through VIIIA in the older convention tell you how many valence electrons an element has. The valence electrons are the outer electrons that are most important in determining the chemical bonding and other properties of the element.

The horizontal rows of the table are called *periods*, and are numbered 1 through 7 starting with the row that contains only H and He.

The Periodic Table of the Elements

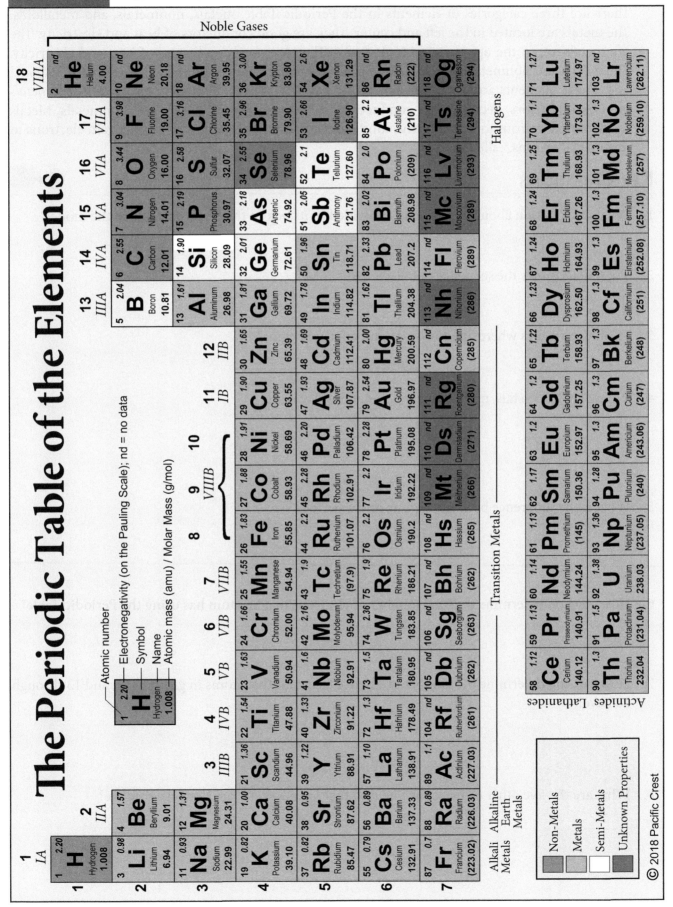

Noble Gases

Halogens

Transition Metals

Lanthanides   Actinides

Alkali Metals
Alkaline Earth Metals

Non-Metals
Metals
Semi-Metals
Unknown Properties

Electronegativity (on the Pauling Scale); nd = no data

Atomic number
Symbol
Name
Atomic mass (amu) / Molar Mass (g/mol)

© 2018 Pacific Crest

INFORMATION

There are three categories of elements in the Periodic Table: metals, nonmetals, and metalloids. The metals are located in the left and center. They are good conductors of heat and electricity. The nonmetals are in the upper right-hand corner. They are poor conductors of heat and electricity. The metals and nonmetals are separated by the *metalloids*, which are seven elements on a diagonal line. These elements are B, Si, Ge, As, Sb, Te, and At. The metalloids are also called *semimetals* or semiconductors, because their conductivity is between that of metals and nonmetals. Metals readily lose electrons to form positive ions, called *cations*, and nonmetals readily gain electrons to form negative ions, called *anions*.

KEY QUESTIONS

1. What information about an element is provided in its box in the Periodic Table in the model?

2. What determines the sequence of the elements?

3. What determines where one row stops and another begins?

4. Where are the metals, nonmetals, and metalloids located?

5. What is the difference between a group and a period?

6. How can you determine the total number of electrons that an atom has using the Periodic Table?

7. How can you determine the number of valence electrons that atoms in groups 1, 2, and 13 through 18 have?

8. What are the five other elements like helium that are gases and are not very reactive?

1. Write the name, symbol, atomic number, and average mass for the Group 2 metal in Period 5.

2. Write the name, symbol, average mass, and number of protons for the Group 16 nonmetal in Period 3.

3. Write the name, symbol, average mass, and number of electrons for the Group 14 metaloid in Period 4.

4. Write the name and symbol of the element that has 45 electrons.

5. Name two elements that have properties similar to potassium, K. How many valence electrons to these elements have?

6. Name two elements that have properties similar to chlorine, Cl. How many valence electrons do each of these three elements have?

7. Using atomic symbols, list the elements in Period 3 in order of increasing number of electrons.

8. Using atomic symbols, list the elements in Group 16 in order of increasing number of protons.

9. Using atomic symbols, list the elements in Group 11 in order of increasing atomic mass.

10. What element comes after lanthanum on the Periodic Table? Where is it shown in **Model 1**?

11. What element comes after lutetium on the Periodic Table? Where is it shown in **Model 1**?

ACTIVITY 03-1 Nomenclature: Naming Compounds

WHY?

The term *nomenclature* refers to a system of principles, procedures, and terms for naming things. Names are an intrinsic part of the way we communicate information. So in order to talk about chemical compounds, which are substances formed from two or more different elements, in a meaningful way, they need to have names that that tell us something about their composition or molecular formula.

LEARNING OBJECTIVE

- Understand and apply the nomenclature for binary ionic compounds, binary covalent compounds, strong acids, and polyatomic ions

SUCCESS CRITERIA

- Correctly write the molecular formula of a compound when given its name
- Correctly name a compound when given its molecular formula

PREREQUISITE

02-3 *The Periodic Table of the Elements*

INFORMATION

Chemical compounds can be classified according to their structure and bonding; the system of nomenclature reflects these distinctions. Two major classes of compounds can be identified by the types of elements that compose them and the resulting bonding that is involved. These two classes are *ionic compounds*, which are composed of metals and non-metals such as sodium chloride ($NaCl$) and *covalent compounds*, which are composed of non-metals with other non-metals, such as water (H_2O) or carbon dioxide (CO_2). These simple classifications are easy to use for *binary compounds* which contain only two elements. For covalent compounds, the number of atoms of any two given elements that can combine varies, so the name must identify the numbers of atoms of each type in the formula using the numerical prefixes, *mono, di, tri, tetra*, etc. See your textbook or the additional resources for all of these and memorize them. *Mono* is usually omitted for the first element written, which is the more metallic (less electronegative) one.

MODEL 1   NAMES OF BINARY COVALENT COMPOUNDS

Molecular Formula	Name	Molecular Formula	Name
HCl	hydrogen chloride	CCl_4	carbon tetrachloride
H_2S	dihydrogen sulfide	N_2O_5	dinitrogen pentoxide
CO	carbon monoxide	NO_2	nitrogen dioxide
CO_2	carbon dioxide	SF_6	sulfur hexafluoride
PBr_3	phosphorous tribromide		

KEY QUESTIONS

1. What does the term *binary covalent* mean?

2. In both the molecular formula of a compound and its name, which element is given first: the more metallic or the less metallic one?

3. What ending is applied to the root of the more electronegative element in the name of a compound?

4. What prefixes are used to indicate the number of atoms when there is more than one possibility?

MODEL 2   **NAMES OF BINARY IONIC COMPOUNDS**

Molecular Formula	Name	Molecular Formula	Name
NaCl	sodium chloride	$CoCl_2$	cobalt(II) chloride
KBr	potassium bromide	$CoCl_3$	cobalt(III) chloride
$MgCl_2$	magnesium chloride	FeO	iron(II) oxide
$CaCl_2$	calcium chloride	Fe_2O_3	iron(III) oxide
CaO	calcium oxide	Al_2O_3	aluminum oxide

KEY QUESTIONS

5. In both the molecular formula and the name of a binary ionic compound, which element is given first: the metal or the nonmetal?

6. What ending is applied to the root of the nonmetal in an ionic compound?

7. When a metal ion can form more than one kind of cation (an ion with a positive charge), how is the charge on the cation indicated?

8. Since a Roman numeral is not included in the names of the alkali metal and alkaline earth metal compounds, how do you determine the charge on the cation and the number of anions associated with it?

9. What are some similarities and differences in the names of binary covalent and binary ionic compounds?

EXERCISES

1. Use your textbook or other reference resource to make a list of some binary covalent compounds that require the use of number prefixes in the name (mono, di, tri, tetra, penta, and hexa).

2. Use your textbook or other reference resource to make a list of metals that require the use of Roman numerals to specify the charge on the cation. Be sure to include the charges that those metal cations commonly have.

3. For the following compounds, identify the compound as **covalent** or **ionic** and complete the table by filling in the missing entries.

Type (ionic or covalent)	Molecular Formula	Name
	CS_2	
	N_2O_4	
		sodium sulfide
	Cu_2O	
	$CuCl_2$	

MODEL 3   NAMES OF ACIDS

Acids and bases are extremely important chemical compounds. You will learn more about them later. For now you need to be able to recognize the names of some common acids and the information they contain. Note that common binary acids can go by either their molecular name or their acid name; for example, HCl can be referred to as either hydrogen chloride or hydrochloric acid.

Molecular Formula	Name
HCl	hydrochloric acid
HOCl	hypochlorous acid
$HClO_2$	chlorous acid
$HClO_3$	chloric acid
$HClO_4$	perchloric acid

Molecular Formula	Name
H_2SO_3	sulfurous acid
H_2SO_4	sulfuric acid
HNO_3	nitric acid

KEY QUESTIONS

10. When two similar acids exist with different numbers of oxygen atoms, how do the names differ?

11. When four similar acids exist with different numbers of oxygen atoms, how do the names distinguish between them? Identify the names and formulas for the four acids containing the elements bromine and oxygen.

12. What are the names of the acids in **Model 3** that have one hydrogen atom?

13. What are the names of the acids in **Model 3** that have two hydrogen atoms?

MODEL 4	NAMES OF POLYATOMIC IONS

Molecular Formula	Name	Molecular Formula	Name
SO_4^{2-}	sulfate	CO_3^{2-}	carbonate
SO_3^{2-}	sulfite	NH_4^+	ammonium
NO_3^-	nitrate	H_3O^+	hydronium
NO_2^-	nitrite	OH^-	hydroxide

KEY QUESTIONS

14. When two similar polyatomic ions exist with different numbers of oxygen atoms, how do the names distinguish between these?

15. Identify the oxo-acid and the polyatomic ion that both contain four oxygen atoms and one sulfur atom. What is the relationship between the number of hydrogen atoms in the acid and the ionic charge of the polyatomic ion for this pair?

16. How does the name of a polyatomic ion indicate that it has a positive charge?

EXERCISE

4. Some of the acids listed in **Model 3** can be formed from a polyatomic ion with a similar name. Write the names and the molecular formulas of each polyatomic ion and its corresponding acid below. Identify how the acid is formed from the polyatomic ion in each case.

WHY?

Molecular structures can be rather complex, but often, repeating structural patterns allow chemists to make generalizations that help simplify our understanding of these structures. Later activities will cover methods of predicting the positions of atoms and bonds in molecules in much more detail; for now, we will examine simple molecular representations using *structural formulas* and *Lewis structures*. Patterns of bonding in organic molecules based on carbon-carbon bonds can also be represented very quickly, by using *line drawings* to represent bonds between carbon atoms. These representations serve as a useful shorthand notation for organic structures.

LEARNING OBJECTIVE

- Represent molecules using molecular formulas, structural formulas, Lewis Structures, and line drawings

SUCCESS CRITERION

- Convert between molecular formulas, structural formulas, Lewis structures, and line drawings

PREREQUISITE

03-1   *Nomenclature: Naming Compounds*

INFORMATION

Molecular formulas that simply list the number of atoms covalently bonded in a molecule don't provide much information about the arrangement of the atoms and the bonds between them. However, it is possible to predict the arrangements of some simple inorganic molecules and polyatomic ions based on how their molecular formulas are written.

In many molecules, a central atom will be bonded to several other atoms that surround it. Since hydrogen, in nearly all cases, is only capable of forming one bond, it will not be the central atom in a molecule. It is often written first in a molecular formula, in order to indicate that it can be lost as H^+ (by an acid such as nitric acid, HNO_3). In *oxo-acids* (acids that contain oxygen), the hydrogen atom is usually bonded to one of the oxygen atoms, which is then bonded to the central atom. In the case of nitric acid, the central atom is nitrogen.

Otherwise, the central atom is usually written first and is is found lower and further to the left on the periodic table. And example is CCl_4, which has carbon as the central atom and four chlorine atoms bonded to it. In later activities (Chapter 8), you will learn how to predict, using *Lewis structures*, how many electrons make up particular covalent bonds. These diagrams show both the covalent bonds, as lines between atoms, and non-bonding electrons, as dots.

Structural formulas for simple molecules are written in a way that indicates the relative positions of the atoms involved. For example, while hydrogen peroxide could be written H_2O_2, the arrangement HOOH illustrates that the two oxygen atoms are bonded to each other, and each of the hydrogen atoms is bonded to one oxygen.

Based on content contributed by Joseph Lauher, Stony Brook University,
modified by David Hanson, Stony Brook University and John Goodwin, Coastal Carolina University

MOLECULAR FORMULAS AND LEWIS STRUCTURES FOR INORGANIC MOLECULES

Name	Molecular Formula	Lewis Structure
sulfur hexafluoride	SF_6	
sulfuric acid	H_2SO_4	
nitrite ion	NO_2^-	
ammonium ion	NH_4^+	

KEY QUESTIONS

1. How is the central atom in each molecule or ion indicated in the molecular formula?

2. Which molecule or molecules in **Model 1** have only single bonds?

3. How are non-bonding electrons shown in the Lewis structure?

4. What feature of the Lewis structure identifies polyatomic ions?

5. Is any information about the three-dimensional molecular structure given by the Lewis structure?

INFORMATION

While simple molecular formulas usually do not contain much information about molecular structure, carbon-based molecules are a different matter. Predictable C-C bonding and carbon's strong tendency to form a total of four bonds sometimes allow molecular structure to be deduced from a written formula. However, Lewis structures are necessary to show details like branched carbon chains and groups that contain elements besides carbon and hydrogen.

A *line drawing* is a special type of structural formula short-hand, wherein carbon atoms and hydrogen atoms with which they share a bond are *not* shown. Lines are used to represent carbon-carbon bonds with the approximate bond angles between carbon atoms shown. The appropriate number of hydrogen atoms to complete the total of four bonds for each carbon atom is generally only implied. Line drawings are often more convenient than Lewis structures because they provide some information about molecular geometry and are easier to draw. If you go on to study organic chemistry, you will use line drawings extensively.

MODEL 2	REPRESENTATIONS OF ORGANIC MOLECULES	

Name Molecular Formula Structural Formula	Lewis Structure	Line Drawing
ethane C_2H_6 CH_3CH_3	H—C—C—H (with H above and below each C)	—
ethene (aka ethylene) C_2H_4 CH_2CH_2	C=C (with two H on each C)	=
propane C_3H_8 $CH_3CH_2CH_3$	H—C—C—C—H (with H above and below each C)	⌃

Name Molecular Formula Structural Formula	Lewis Structure	Line Drawing
butane C_4H_{10} $CH_3CH_2CH_2CH_3$		
ethanol C_2H_6O CH_3CH_2OH		
methyl amine CH_5N CH_3NH_2		

KEY QUESTIONS

1. What is the Lewis structure for ethene?

2. What does the line drawing of ethene look like?

3. What do the lines in Lewis structures and line drawings represent?

4. What element is assumed to be present at the end of a line in a line drawing, even though no atomic symbol is written?

5. What bonds and elements are missing entirely from the line drawing of ethene?

6. How many bonds does each carbon atom in a molecule of ethene form, according to its Lewis structure?

7. How can one determine the number of hydrogen atoms bound to a given carbon atom from a line drawing?

8. Why is the line drawing of propane bent?

9. How many bonds are formed by the central carbon atom in propane, according to its line drawing? (How many are implied, rather than explicitly shown?)

EXAMPLE

Consider the line structure of the molecule 1-butene:

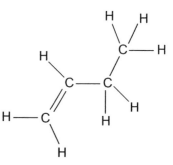

Starting from the left, the first carbon has two bonds, so it must also have two hydrogen atoms. The second carbon has three bonds, so it has one hydrogen atom. The third carbon has two bonds and thus two hydrogen atoms, and the fourth and final carbon has only one bond and therefore must have three hydrogen atoms. If we sum all of the necessary hydrogen atoms, we get the molecular formula C_4H_8, and the structural formula is shown at right:

Chapter 3: Molecules and Compounds

In the space provided below, write both the molecular and structural formulas for each of the following line drawings. Note that the structural formula explicitly includes all carbon and hydrogen atoms.

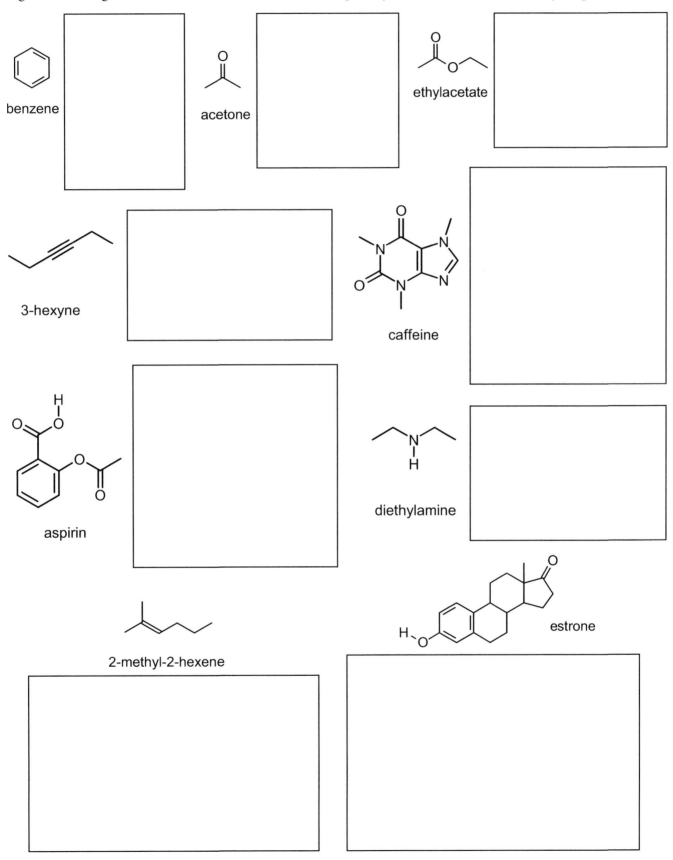

benzene

acetone

ethylacetate

3-hexyne

caffeine

aspirin

diethylamine

2-methyl-2-hexene

estrone

Moles and Molar Mass

WHY?

To keep track of the huge numbers of atoms and molecules in samples that are large enough to see, chemists have established a unit of counting called the *mole* (abbreviated mol) and a unit of measure called the *molar mass*, which has units of g/mol. By using the idea of a *mole* and *molar mass*, you will be able to count out specific numbers of atoms or molecules simply by weighing them. The system allows convenient scaling up from the atomic level, with masses based on atomic mass units (amu), to the laboratory level using the same numerical values, but in g/mole. This ability to use correct atomic and mole ratios is essential in conducting research in chemistry and biology, and applying chemistry in technology and the health sciences.

LEARNING OBJECTIVES

• Understand the relationship between the mole and Avogadro's number

• Understand the meaning of the molar mass of a substance

• Recognize that the molar mass is an average of all the isotopic masses of an element

SUCCESS CRITERIA

• Quickly convert between the number of atoms, moles, and the mass of a sample by using Avogadro's number and the molar mass appropriately

• Calculate the molar mass from isotopic abundances and isotopic masses

PREREQUISITES

01-2 *Unit Analysis*

01-3 *Significant Figures in Data*

02-1 *Atoms, Isotopes, and Ions*

MODEL 1   A MOLE IS A COUNTING UNIT

1 **pair** of objects	1 **dozen** objects	1 **gross** of objects	1 **mole** of objects
2 objects	**12** objects	**144** objects	**6.02214 × 10²³** objects

KEY QUESTIONS

1. How many kittens are there in a dozen kittens?

2. How many kittens are there in a gross of kittens?

3. How many kittens are there in a mole of kittens?

4. How many atoms are there in a dozen atoms?

5. How many atoms are there in a gross of atoms?

6. How many atoms are there in a mole of atoms?

7. In what way are the meanings of the terms *pair, dozen, gross,* and *mole* similar?

How do their meanings differ?

INFORMATION

The number of objects in a mole (6.02214×10^{23}) is so important in chemistry that it is given a name. It is called *Avogadro's number,* and it has units of objects /mol.

Avogadro's number is determined by the number of carbon atoms in exactly 12 g of pure carbon-12 isotope. Remember that the mass of one carbon-12 atom is also defined as exactly 12 amu. The mole system is set up so that the same numerical values can be used in atomic mass units for single atoms and molecules or grams/mole for moles of atoms and molecules of the same kind. So, the masses of the elements on the Periodic Table refer to atoms in amu or moles of atoms in grams. The *molar mass* is the mass of a mole of objects. It has units of g/mol.

$1 \text{ amu} = 1.66054 \times 10^{-24} \text{ g}$

EXERCISES

1. A single carbon-12 atom has a mass of 12 amu by definition of the atomic mass unit. Convert 12 amu to grams, and then calculate the mass in grams of a mole of carbon-12 atoms.

2. A single oxygen-16 atom has a mass of 15.9949 amu. Convert this mass to grams, and then calculate the mass in grams of a mole of oxygen-16 atoms.

3. Based on your results for Exercises 1 and 2, identify the relationship between the numerical values of the mass of an atom in amu and its molar mass in g/mol.

4. Sometimes the mole is called the "chemist's dozen". Explain what is implied by that.

| **MODEL 2** | **BEADS IN A JAR — THE AVERAGE MASS OF A MIXTURE OF OBJECTS** |

Natural samples of most elements are mixtures of different isotopes. The mass of Avogadro's number of atoms in such a sample is not the molar mass of a single isotope, but is rather an abundance-weighted average of the masses of all the isotopes for that element. Exploration of **Model 2** will guide you in determining these average molar masses.

Table 1

Bead Color	Mass of a Single Bead	Number in the Jar	Percent Abundance
red	2.0 g	50	50%
blue	2.5 g	30	30%
yellow	3.0 g	20	20%

KEY QUESTIONS

8. Which color bead has the largest mass?

9. Which color bead is present in the largest number?

10. Is the average mass of a bead in the jar equal to 2.5 g, which is [(2.0 + 2.5 + 3.0) ÷ 3]? Explain.

11. Is the average mass of a bead in the jar greater than or less than 2.5 g? Explain, without performing a calculation; just examine the information in Table 1.

12. How can you calculate the *weighted average mass* of a bead in the jar using the percent abundance given in Table 1? Provide an explanation, then perform the calculation.

13. How is the calculation of weighted average mass similar to the calculation of your course grade in chemistry class?

EXERCISES

5. Using your calculation of the weighted average mass of beads as a guide, explain how to determine the molar mass of naturally-occurring boron from the data given in Table 2 below. There are only two naturally-occurring isotopes.

Table 2

Isotope	Isotope Mass (amu or g/mol)	Percent Abundance
boron-10	10.0129	Unknown
boron-11	11.0093	80.22%

6. Compare the number you calculated for the molar mass of boron in Exercise 5 with the number given below the symbol for boron in the Periodic Table. From this comparison, identify the information that is provided by the numbers just below the atomic symbols in the Periodic Table.

7. Distinguish between the following: *isotope mass (amu)*, *isotope mass (g/mole)*, *mass reported on the Periodic Table (amu and g/mole)*, and *mass number*. Does any specific atom have the average mass (in amu) given on the periodic table?

8. Calculate the number of atoms in exactly 2 moles of naturally-occurring helium.

9. Calculate the number of moles corresponding to 2.007×10^{23} atoms of helium.

10. Calculate the mass in grams of 2.5 moles of argon. The molar mass of argon is 39.95 g/mol.

11. Calculate the number of moles in 75 g of iron. The molar mass of iron is 55.85 g/mol.

12. Calculate the number of atoms in 0.25 moles of uranium.

13. Calculate the mass of 12.04×10^{23} atoms of uranium. The molar mass of uranium is 238.0 g/mol.

Got It!

1. Identify the statement below that is correct and explain why it is correct:

   a. The molar mass of an element divided by Avogadro's number gives the average mass of an atom of that element in grams.

   b. The molar mass of an element divided by Avogadro's number gives the mass of one atom of that element in grams.

2. If you have 1 g samples of several different elements, will the sample with the largest or smallest molar mass contain the fewest atoms? Explain.

3. Write the units that result from the following mathematical operations.

   a. number of objects / Avogadro's number =

   b. moles × Avogadro's number =

   c. mass / molar mass =

   d. moles × molar mass =

PROBLEMS

1. The atomic mass of ^{35}Cl is 34.971 amu, and the atomic mass of ^{37}Cl is 36.970 amu. In a natural sample, 75.77% of the atoms are ^{35}Cl, and the rest are ^{37}Cl. Describe how you can calculate the molar mass of chlorine (which is the weighted average mass) from these data, then calculate a value for the molar mass of chlorine.

2. A mass of 32.0 g of molecular oxygen reacts completely with 6.02214×10^{23} atoms of carbon.

    a. What is the ratio of moles of C to moles of O in the product?

    b. Given that mass is conserved in a chemical reaction, what is the mass of the product produced?

    c. Is the product carbon monoxide, CO, or carbon dioxide, CO_2?

03-4 Determination of Chemical Formulas

WHY?

Chemical formulas tell you how many atoms of each element are present in a compound. Molecular formulas are used for covalent molecules and specify the total number of atoms of each element in the molecule. Ionic compounds, which are composed of specific whole-number ratios of cations and anions in alternating three-dimensional lattices instead of molecules, are better described by empirical formulas. For an ionic compound, the empirical formula is the simplest whole-number ratio of ions contained in the lattice; for a covalent compound, it is the simplest whole-number ratio of atoms contained in a sample. For both ionic and covalent compounds, empirical formulas can be determined by measuring the mass of each element present in a sample of the compound. A molecular formula can be found from the empirical formula if the molar mass of the compound is known. This conversion of macroscopic quantities of material (grams) to the microscopic composition (number of atoms of each element present in a molecule) is used as a first step in determining the structure of the substance at the atomic level.

LEARNING OBJECTIVE

- Understand the relationship between the mass percent composition of a chemical compound, its molar mass, and its empirical formula, and how this relates to molecular formula

SUCCESS CRITERIA

- Quickly calculate mass percent composition from a molecular or empirical formula
- Determine the empirical formula of a compound from its mass percent composition
- Determine the molecular formula of a compound from its empirical formula and molar mass

PREREQUISITE

03-3 *Moles and Molar Mass*

INFORMATION

The molecular or chemical formula of a molecule tells you the number of atoms of each element that comprise the molecule. It also tells you the number of moles of each element needed to make 1 mole of the compound. For example, the molecular formula for glucose (sugar), $C_6H_{12}O_6$, tells you that each molecule contains 6 carbon atoms, 12 hydrogen atoms, and 6 oxygen atoms, and that the same numbers of moles of each element are needed to make 1 mole of glucose.

The percent composition by mass of a compound provides information that is useful in determining molecular formulas. The mass percent of an element in a sample of a pure compound is calculated in the following way:

$$\text{mass \% of an element} = \frac{\text{mass of the element present in a sample}}{\text{total mass of the sample}} \times 100\%$$

The mass percent of an element in a compound can also be calculated if the molecular formula is known. This calculation is done in the following way:

$$\text{mass \% of an element} = \frac{\text{mass of the element in one mole of the compound}}{\text{molar mass of the compound}} \times 100\%$$

For example, to calculate the percent carbon in propene, C_3H_6: the mass of carbon in 1 mol of propene is 3×12.011 g/mol = 36.033 g/mol. The molar mass of propene is 42.080 g/mol, so dividing 36.033 by 42.080 and multiplying by 100% produces 85.63% carbon.

The empirical formula is the simplest whole-number ratio of atoms in a compound. For the examples just discussed, the empirical formula of glucose is obtained by dividing all the subscripts by six to give CH_2O. For propene, the empirical formula is CH_2. If we consider the common ionic compound sodium chloride, which is composed of equal numbers of Na^+ cations and Cl^- anions, the empirical formula NaCl is completely sufficient. *Empirical* means observed, so an empirical formula is the simplest experimentally observed ratio of elements in a compound. The methods differ, but essentially all are based on separating the elements in the compound from each other by chemical reactions, isolating the products and then finding the masses of the separated elements in the compound. Using molar masses, we can then determine the empirical formula, which is the simplest mole ratio of elements.

MODEL 1   CHEMICAL ANALYSIS OF ACETIC ACID

Acetic acid is the active ingredient in vinegar. A chemical analysis of 157.5 g of acetic acid provided the following information:

Element	Mass of Element (g)	Moles of Element	Whole Number Ratio
carbon	63.00	5.246	5.246 / 5.246 = 1
oxygen	83.93	5.246	5.246 / 5.246 = 1
hydrogen	10.57	10.486	10.486 / 5.246 = 2

moles = mass of element / molar mass of element

The 1:1:2 ratio in the last column means that the empirical formula is (COH_2).

KEY QUESTIONS

1. How was the number of moles of each element in the acetic acid sample calculated?

2. How were the ratios of the elements in acetic acid determined from the moles of each element present in the sample?

3. How is the empirical formula determined from the ratios of the elements present in the sample?

4. What information does the empirical formula provide?

5. What is the relationship between the ratio of moles of each element present in the sample and the ratio of the number of atoms of each element present in each molecule of acetic acid?

6. Why is the atomic mass percent composition of an unknown chemical compound an important quantity to determine in a chemical analysis?

EXERCISES

1. Sodium carbonate has the molecular formula Na_2CO_3. Calculate the molar mass of this compound, the mass percent composition of each element in the compound, and the mass of each element present in a 57.2 g sample.

2. A sample of baking soda was found to consist of 4.561 g Na, 0.2000 g H, 2.383 g C, and 9.520 g O. Calculate the moles of each element present, and determine the empirical formula of baking soda.

PROBLEMS

1. Show how you can determine the molecular formula for acetic acid from the empirical formula CH_2O and the added information that the molar mass of acetic acid is 60.05 g/mole.

2. One of the chlorofluorocarbons (a freon), which is used in refrigerator compressors (and which contributes to destruction of ozone in the upper atmosphere), has a molar mass of 132.9 g/mole and a percent composition of 53.34% Cl, 28.59% F, and 18.07% C. Hint: Whenever the data give the percent of the elements present rather than the mass, you can assume any sample size and calculate the mass. It is particularly convenient to assume the data came from a 100 g sample; explore why.

   a. How many grams of chlorine are there in a 100 g sample of the freon?

   b. How many moles of chlorine are there in a 100 g sample of the freon?

   c. What is the ratio of moles of chlorine to moles of carbon in this freon?

   d. Determine the empirical and molecular formulas of this freon.

3. Combustion of 10.68 mg of a compound containing only C, H, and O produces 16.01 mg CO_2 and 4.37 mg H_2O. All of the carbon in the sample is trapped as carbon dioxide and all of the hydrogen in the sample is trapped as water.

   a. Find the mass of the carbon in the carbon dioxide using the conversion factor based on the molar masses of carbon and carbon dioxide (i.e., there are 12.01 g C in the sample for every 44.01 g CO_2 produced).

   b. Perform a similar calculation for the mass of hydrogen in the water, again based on molar mass ratios and keeping in mind there are two moles of hydrogen in one mole of water.

   c. Now that you have the masses of carbon and hydrogen in the sample, find the mass of the oxygen in the sample by finding the difference in the total sample mass and the sum of the carbon and hydrogen mass, since oxygen is the only other element present.

d. Find the empirical formula from these masses (as you did before).

e. The molar mass of the compound is 176.1 g/mol. What is the molecular formula of the compound?

WHY?

Chemical reaction equations are fundamental tools for communicating how chemical reactions and changes in physical state occur. Using the mole concept, chemical reaction equations represent the quantitative relationships between the numbers of reactant and product species involved at both the atomic level and the laboratory level. To make use of these equations, you first need to be able to write formulas for chemical compounds, balance the chemical equations to account for all the atoms and ionic charge, and deduce information from equations about the amounts of material needed for and produced by the reaction.

LEARNING OBJECTIVE

- Understand and make use of chemical reaction equations

SUCCESS CRITERIA

- Write and balance chemical reaction equations
- Determine the amounts of substances consumed and produced in a chemical reaction

PREREQUISITES

03-1 *Nomenclature: Naming Compounds*

03-3 *Moles and Molar Mass*

MODEL 1 BALANCING THE COMBUSTION OF PROPANE

Propane is a combustible hydrocarbon gas used in gas grills and in gas logs in fireplaces. It has the molecular formula C_3H_8. Combustion reactions with hydrocarbons should be familiar to you: also known as burning, they take place when a hydrocarbon reacts with O_2, usually from the air, and produces carbon dioxide (CO_2) and water (H_2O). All chemical reaction equations must show the correct chemical formulas of all the reactants and products and, additionally, account for the total number of each kind of atom on both sides of the equation. The reactants that are present before the reaction takes place are shown on the left side of the single arrow and the products of the reaction on the right side of the arrow. The physical states of each reactant and product are usually shown in parentheses, with (s), (l), (g), and (aq) being used to show solids, liquids, gases, and species in aqueous solution. The skeletal reaction, which shows an **unbalanced** equation but uses correct formulas for reactants and products, for the combustion of propane is:

$$C_3H_8 + O_2 \rightarrow CO_2 + H_2O$$

To balance the reaction, start by creating a list of each element present in the reactants and products and the number of atoms of each element in the reaction:

Reactants	$\rightarrow$	Products
$C_3H_8 + O_2$	$\rightarrow$	$CO_2 + H_2O$

Atom	How many?
C	3
H	8
O	2

Atom	How many?
C	1
H	2
O	3

Continue to balance the reaction by finding the more complicated formulas first, and then add balancing coefficients to indicate the number of reactants or products necessary.

 Do *not* change the compound formulas to balance reactions! NEVER do anything like changing H_2O to HO_2 just because you want two oxygen atoms in order to balance an equation.

In propane, there are **three carbon atoms**, indicating there must be **three carbon dioxide molecules produced**.

$$C_3H_8 + O_2 \rightarrow 3CO_2 + 4H_2O$$

There are also **eight hydrogen atoms** in propane, indicating there must be **four water molecules** produced, since **four molecules times two hydrogen atoms per molecule gives eight hydrogen atoms**.

Now that we've balanced the carbons and hydrogens on each side of the equation, the new tally of atoms is:

Reactants	→	Products
$C_3H_8 + O_2$	→	$3CO_2 + 4H_2O$

Atom	How many?
C	3
H	8
O	2

Atom	How many?
C	3
H	8
O	10

Oxygen is the easiest to balance since it is a single element (rather than a compound), so saving it for last makes sense. The 10 oxygen atoms present among the products mean that we need to place a balancing coefficient of **5** in front of O_2 on the left side. Addition of the physical states completes the job of balancing this reaction. Note that the order in which the products and reactants are written doesn't matter.

$$C_3H_8(g) + 5O_2(g) \rightarrow 3CO_2(g) + 4H_2O(g)$$

The final tally of atoms is then:

Reactants	→	Products
$C_3H_8 + 5O_2$	→	$3CO_2 + 4H_2O$

Atom	How many?
C	3
H	8
O	10

Atom	How many?
C	3
H	8
O	10

KEY QUESTIONS

1. What are the reactants in the combustion of propane?

2. What are the products in the combustion of propane?

3. What meaning is given to the arrow in a chemical reaction equation?

4. In a reaction equation, what information is provided by the numerical subscripts in the molecular formulas for the reactants and products?

5. In a balanced reaction equation, what information is provided by the stoichiometric coefficients? Note: The number in front of a reactant or product is called a *stoichiometric coefficient*. The absence of an explicit stoichiometric coefficient means the value is 1.

6. What process is used to determine the total number of atoms of each element in a chemical reaction, using formula subscripts and balancing coefficients?

7. Why is it possible to interpret a reaction equation in each of the following two ways?
    a. As specifying how many molecules of reactants are consumed to produce the specified number of product molecules.
    b. As specifying how many moles of reactants are consumed in producing the indicated number of moles of products.

EXERCISES

1. Write balanced reaction equations for the following reactions. Be sure to add states if they are missing.

   a. $Na_2CO_3(s) + HCl(aq) \rightarrow NaCl(aq) + CO_2(g) + H_2O(l)$

   b. $CH_3OH(l) + O_2 \rightarrow CO_2(g) + H_2O(g)$

   c. In the gas phase: $N_2 + H_2 \rightarrow NH_3$

   d. $LiOH(s) + CO_2 \rightarrow Li_2CO_3(s) + H_2O(l)$

   e. $Pb(NO_3)_2 (aq) + H_2S(g) \rightarrow PbS(s) + HNO_3(aq)$

| MODEL 2 | BALANCING COEFFICIENTS IN REACTION STOICHIOMETRY |

The balanced chemical reaction equation for the combustion of propane from **Model 1** describes the numerical relationship between the number of molecules of reactants and products on the molecular level and, by using the mole concept, on the laboratory level as well (where the balancing coefficients indicated numbers of moles.)

$$C_3H_8(g) + 5O_2(g) \rightarrow 3CO_2(g) + 4H_2O(g)$$

The reaction can be written in words as, "One mole of propane gas reacts with five moles of oxygen gas to produce three moles of carbon dioxide gas and four moles of water vapor."

The relationships between the moles of reactants and products given by the balancing coefficients can be used to create conversion factors, sometimes called *mole ratios*. For example, we can determine the number of moles of oxygen that would react with 0.50 mol propane and the number of moles of carbon dioxide that would be produced by starting with 0.50 mol propane, then creating the required conversion factors:

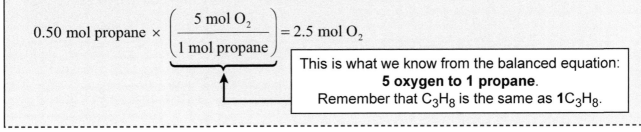

$$0.50 \text{ mol propane} \times \left(\frac{5 \text{ mol } O_2}{1 \text{ mol propane}} \right) = 2.5 \text{ mol } O_2$$

This is what we know from the balanced equation: **5 oxygen to 1 propane**. Remember that C_3H_8 is the same as **1C_3H_8**.

The conversion factor implies that for every 1 mol of propane burned, 5 mol of O_2 are required.

Likewise for carbon dioxide:

$$0.50 \text{ mol propane } \times \left(\frac{3 \text{ mol } CO_2}{1 \text{ mol propane}} \right) = 1.5 \text{ mol } CO_2$$

The conversion factor implies that for every 1 mol of propane burned, 3 mol of CO_2 are produced.

ANY combination of reactants and products can be treated in this way.

Key Questions

8. When using balancing coefficients to create mole ratio conversion factors, why is it insufficient to simply use moles? What else is required?

9. If you were given an amount of reactant propane in grams instead of moles and needed to calculate grams of carbon dioxide produced, what additional information and calculation steps would be required to obtain the answer?

Exercises

2. Using the reaction equation for the combustion of propane, determine the number of grams of oxygen that would react with 52 g of propane and the number of grams of water that would be produced.

As you begin to study chemical reaction types involving ions in solution, you will encounter many reactions requiring balancing that can be greatly simplified if you recognize polyatomic ions that retain their essential covalently-bonded structures during the conversion of reactants to products. You will soon encounter net ionic equations that do not show all of the ions in solution in the balanced reaction equation, instead showing only those that react. It is necessary to balance ionic charge in these reactions. Here are a few examples of reactions involving polyatomic ions.

Ionic dissociation of soluble ionic compounds produces dissolved ions, here the reactants and products include the carbonate ion. Each side of the reaction has a net ionic charge of zero:

$$Na_2CO_3(aq) \rightarrow 2Na^+(aq) + CO_3^{2-}(aq)$$

Acid ionization produces hydronium ion, H^+, and the conjugate base in water; in this case, the nitrate ion. While ions do not exist in HNO_3 before it reacts with water, balancing the reaction is made easier by recognizing the NO_3^- polyatomic ion as the conjugate base of the acid:

$$HNO_3(aq) \rightarrow H^+(aq) + NO_3^-(aq)$$

In **acid-base neutralization reactions**, protons are transferred from acids to bases. Recognizing the hydroxide ion, OH^-, that can be protonated to make water, makes balancing easier:

$$2HNO_3(aq) + Ba(OH)_2(aq) \rightarrow 2H_2O(l) + Ba(NO_3)_2(aq)$$

In **precipitation reactions** with the carbonate ion, ions combine to form an insoluble solid. The total ionic charge of reactants and products is constant and the polyatomic carbonate ion is unchanged:

$$Mg^{2+}(aq) + CO_3^{2-}(aq) \rightarrow MgCO_3(s)$$

You will need to recognize polyatomic ions like these, remembering their names, formulas, and ionic charges as you begin to study chemical reaction types.

KEY QUESTIONS

10. What makes an ion a polyatomic ion?

11. Each of the reactions from **Model 3** appear below. For each:

  a. **Circle** each polyatomic ion that retains its ionic character during the reaction. Below these polyatomic ions, write their names and formulas including their ionic charges.

  b. **Draw a box around** the reactant or product polyatomic ions that correspond to molecules on the opposite side of the reaction equation. **Underline** the portion of the molecule that remains unchanged as it ionizes. Below these polyatomic ions and molecules, write their names and formulas including their ionic charges.

  i. $Na_2CO_3(aq) \rightarrow 2Na^+(aq) + CO_3^{2-}(aq)$

  ii. $HNO_3(aq) \rightarrow H^+(aq) + NO_3^-(aq)$

  iii. $2HNO_3(aq) + Ba(OH)_2(aq) \rightarrow 2H_2O(l) + Ba(NO_3)_2(aq)$

  iv. $Mg^{2+}(aq) + CO_3^{2-}(aq) \rightarrow MgCO_3(s)$

12. For the balanced reaction between barium chloride and sodium sulfate,

$$BaCl_2(aq) + Na_2SO_4(aq) \rightarrow BaSO_4(s) + 2NaCl(aq)$$

  a. What polyatomic ion(s) is/are present? Is its/their structure changed during the reaction?

  b. Why is it necessary to include a balancing coefficient of 2 for the sodium chloride in the product?

c. How could you have written a balanced reaction equation, based only on the names of these reactants and products?

EXERCISE

3. Aqueous phosphoric acid reacts with aqueous sodium hydroxide to produce aqueous sodium phosphate and liquid water.

a. Identify each polyatomic ion involved with its correct name and formula including any ionic charge.

b. Write the correct formulas for each reactant and product.

c. Now that you have the description of the reaction and the formulas involved, write a balanced reaction equation for the reaction.

WHY?

Reactants are not usually present in the exact amounts required by a balanced chemical reaction equation. In situations where amounts of more than one reactant are combined, the reaction stops when one reactant is completely exhausted, leaving an excess of the other reactant. To predict the outcome of such a reaction, it is necessary to perform calculations to determine which reactant, (called the *limiting reactant*), as well as how much product is made and how much of the excess reactant remains.

LEARNING OBJECTIVES

- Identify reactants that limit the extent of a reaction and the amount of product produced
- Determine the amounts of excess reactants remaining after a reaction

SUCCESS CRITERIA

- Identify limiting and excess reactants
- Calculate the amounts of material produced and amounts of excess reactants

PREREQUISITES

03-1 *Nomenclature: Naming Compounds*

03-3 *Moles and Molar Mass*

04-1 *Balanced Chemical Reaction Equations*

MODEL 1   LIMITING INGREDIENT

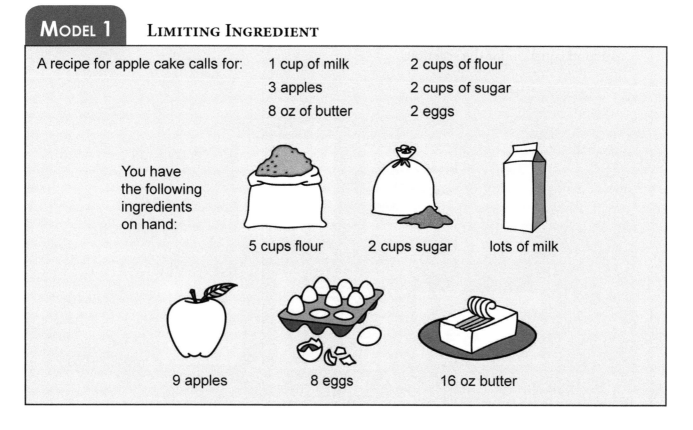

A recipe for apple cake calls for:
  1 cup of milk   2 cups of flour
  3 apples   2 cups of sugar
  8 oz of butter   2 eggs

You have the following ingredients on hand:

5 cups flour   2 cups sugar   lots of milk

9 apples   8 eggs   16 oz butter

1. According to **Model 1**, how much of each ingredient is necessary to make an apple cake?

milk	flour	apples	sugar	butter	eggs

2. How is the recipe for making a cake similar to a chemical reaction equation? How is it different?

3. If you follow the recipe, using only the ingredients on hand in the model, how much of each ingredient will be left over after you have baked the cake?

milk	flour	apples	sugar	butter	eggs

4. Which of the ingredients on hand were in excess of the recipe?

5. Which of the ingredients on hand were consumed completely in making the cake?

6. Which of the ingredients limit or prevent you from making a second, smaller cake?

7. What would be a good definition for the term *limiting ingredient*?

8. Use the cake recipe to create conversion factors for each ingredient. As an example, the conversion factor for milk, showing the ratio of cakes to available cups of milk, is shown below.

milk	flour	apples
$\left(\dfrac{1\,\text{pie}}{1\,\text{cup of milk}} \right)$		

sugar	butter	eggs

9. Use the conversion factors you created to calculate the number of cakes that can be made, given the amount of each ingredient you have on hand.

	milk	flour	apples	sugar	butter	eggs
How many cakes?						

10. Which number of cakes in the calculations in question 9 is the correct one? Why?

1. You want to make 2 dozen standard-sized cookies as specified by a recipe that requires 16 oz of butter, 4 eggs, 3 cups of flour, and 4 cups of sugar. In taking inventory of your supplies, you find that you have 16 oz of butter, 6 eggs, and 3 cups each of flour and sugar.

    a. Express the recipe for these cookies in the form of a reaction equation as started for you below.

    16 oz butter + 4 eggs +

    b. Follow the strategy outlined in the key questions to identify which ingredient will limit the number of cookies you can make.

    c. How many dozen standard-sized cookies can you make?

2. You have 100 bolts, 150 nuts, and 150 washers. You assemble a nut/bolt/washer set using the following recipe or equation.

    $$2 \text{ washers} + 1 \text{ bolt} + 1 \text{ nut} \rightarrow 1 \text{ set}$$

    a. How many sets can you assemble from your supply?

    b. Which is the limiting component? What number of leftover cakeces with there be (if any)?

3. A reaction of hydrogen with oxygen to produce water is described by the following recipe or reaction equation, which says that 2 molecules of hydrogen react with 1 molecule of oxygen to produce 2 molecules of water.

    $$2H_2 + O_2 \rightarrow 2H_2O$$

    You react 120 H_2 molecules with 80 O_2 molecules to produce H_2O. Which is the limiting reactant, hydrogen or oxygen? How many water molecules can you produce from your supply of hydrogen and oxygen? How many left-over unreacted molecules would be left?

MODEL 2 **LIMITING REACTANTS IN MOLES**

A reaction equation can be interpreted in terms of the number of molecules (see Exercise 3) or moles reacting (2 moles of hydrogen react with 1 mole of oxygen to produce 2 moles of water). The only difference is that while a mole can be divided into fractions of a mole, a molecule cannot be divided. Each square below represents one mole of hydrogen, oxygen, and water, respectively.

Figure 1

1 mole of H_2 1 mole of O_2 1 mole of H_2O

When these react according to the reaction equation,

$$2H_2 + O_2 \rightarrow 2H_2O$$

twice as much hydrogen will be used and twice as much water will be produced when compared to the amount of oxygen that reacts. Figure 2 indicates what happens when 1 mole of hydrogen and 1 mole of oxygen are mixed together and react to produce water. It only takes ½ mole of oxygen to react with the 1 mole of hydrogen, so there is ½ mole of oxygen left over.

Figure 2

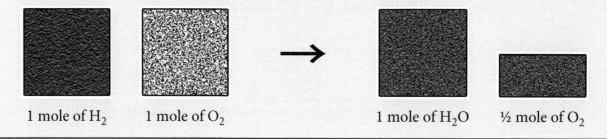

1 mole of H_2 1 mole of O_2 1 mole of H_2O ½ mole of O_2

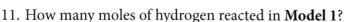

KEY QUESTIONS

11. How many moles of hydrogen reacted in **Model 1**?

12. How many moles of oxygen remained after the reaction?

13. Was hydrogen or oxygen the limiting reactant?

14. In limiting reactant problems, you are sometimes given quantities of reactants in grams. Why do you need to convert these quantities to moles of reactants in order to solve the problem?

4. If 6 moles of hydrogen (H_2) and 4 moles of oxygen (O_2) are mixed and reacted to completion, which is the limiting reactant? How many moles of water would be produced? How much excess reactant remains?

5. If you had 1.73 moles of hydrogen (H_2) and 0.89 moles of oxygen (O_2), which would be the limiting reactant? How many moles of water could you produce from your supply of hydrogen and oxygen? How many moles of excess reactant would remain?

6. If you had 27.3 g of hydrogen and 18.91 g of oxygen, which would be the limiting reactant, and how many grams of water could you produce? How many grams of excess reactant would remain?

PROBLEMS

1. Hydrogen cyanide is used in the production of cyanimid fertilizers. It is produced by the following reaction:

$$2\,CH_4 + 2\,NH_3 + 3\,O_2 \rightarrow 2\,HCN + 6\,H_2O$$

   a. How much hydrogen cyanide can be produced from 100 kg of each of the reactants?

   b. Which is the limiting reactant?

   c. How much excess reactant remains?

2. Aluminum reacts with iodine gas to form aluminum iodide via the following reaction:

$$2Al + 3I_2 \rightarrow 2AlI_3$$

   a. How much aluminum iodide can be produced with 0.70 g of aluminum and 14.5 g of iodine?

   b. What is the limiting reactant?

   c. How much excess reactant remains?

WHY?

You can determine the number of molecules or moles of a solute present from the volume of a solution, provided you know the concentration of the solution. You will encounter problems and situations in this course, in other courses, and in many jobs where you need to know the concentration of solutions. Laundry detergents, medicines, food products, cosmetics, steel, gasoline, and many other materials must contain specific amounts of certain chemical compounds in order to have desired properties.

LEARNING OBJECTIVES

- Understand and use the concept of *molarity*
- Determine how concentration changes upon dilution

SUCCESS CRITERIA

- Accurately calculate solution concentration
- Accurately convert concentration into amount (moles or mass) of material

PREREQUISITE

03-3 *Moles and Molar Mass*

INFORMATION

A solution is formed when one substance is dissolved in another. The substance present in the larger amount is called the *solvent* and the other substance is called the *solute*. Water is a common solvent and solutions involving water as the solvent are called *aqueous solutions* (abbreviated aq).

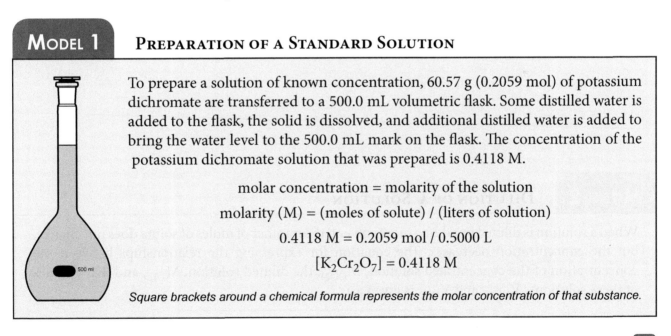

| MODEL 1 | PREPARATION OF A STANDARD SOLUTION |

To prepare a solution of known concentration, 60.57 g (0.2059 mol) of potassium dichromate are transferred to a 500.0 mL volumetric flask. Some distilled water is added to the flask, the solid is dissolved, and additional distilled water is added to bring the water level to the 500.0 mL mark on the flask. The concentration of the potassium dichromate solution that was prepared is 0.4118 M.

molar concentration = molarity of the solution

molarity (M) = (moles of solute) / (liters of solution)

0.4118 M = 0.2059 mol / 0.5000 L

$[K_2Cr_2O_7] = 0.4118$ M

Square brackets around a chemical formula represents the molar concentration of that substance.

KEY QUESTIONS

1. In **Model 1**, what two factors determine the molar concentration of a solution?

2. What unit is used for solution concentration in the model?

3. How is the concentration in units of molarity, i.e., the molar concentration, calculated in the model?

4. What other unit could you use for concentration, other than mol/L?

5. What is a situation that you have encountered where it was important to know a solution concentration?

6. Given the molar concentration of a solution, how can you determine the number of moles of solute in some volume of that solution?

| **MODEL 2** | **DILUTION OF A SOLUTION** |

When a solution is diluted by adding more solvent, the number of moles of solute does not change, but the concentration decreases. The equation for expressing the relationships between the concentration of the concentrated solution, M_{CON}, the diluted solution, M_{DIL}, and the volumes of these solutions, V_{CON}, and V_{DIL}, is given as:

$$M_{CON} \times V_{CON} = \text{moles solute} = M_{DIL} \times V_{DIL}$$

In this equation, most often the units of molarity (mol/L) are used for concentration, the units of L for volume, and the units of moles for the quantity of solute. If the volume is expressed in mL, the amount of solute would be calculated as mmol.

7. The number of moles of solute present doesn't change when the solution is diluted. How can you use this fact, if you are given the initial molar concentration and volume of a solution, to determine the molar concentration of the diluted solution?

8. Rearrange the equation in **Model 2** to find the volume needed to make 500 mL of a dilute solution with a concentration of 5.00×10^{-2} M HCl from a concentrated solution that is 1.00 M HCl. What is the total number of mmoles of HCl present in the dilute solution? What is that concentration in moles?

EXERCISES

1. Calculate the molarity of a 0.175 L sugar solution that was prepared with 0.15 mole of sugar.

2. Determine the volume of 0.235 M sugar solution that can be prepared with 0.470 moles of sugar.

3. A salt solution is to be added to a marine aquarium. Calculate the molarity of a salt solution that is prepared by adding water to 18.65 g of NaCl to give a final volume of 250.0 mL.

4. The solution in Exercise 3 was diluted to a volume of 1.000 L. Calculate the molar concentration of the final solution.

PROBLEMS

1. Calculate the volume of a 4.35 M NaOH (aq) solution that should be used to prepare 275 mL of 0.350 M NaOH (aq).

2. Toxic chemicals in drinking water usually are reported and a safety level specified in units of parts per million (ppm) by mass. What is the molar concentration of arsenic in a water sample that has 1 ppm arsenic (As)? Hint: For aqueous solutions of low concentration like this one, solution density can be approximated as 1.00 g/mL. Why is this acceptable and why is it relevant?

Solving Solution Stoichiometry Problems

Why?

Balanced chemical reactions may be used to determine the relationships between amounts of reactants and products by using the mole ratios given by the balancing coefficients. As you've already seen, these amounts can be given directly as moles or as masses in grams or other units, or as number of atoms, ions or molecules. Since the number of moles of solution-phase reactants and products is determined by concentration and volume, it is also possible to extend stoichiometry concepts to reactions that take place in solution.

Learning Objectives

- Determine the amounts of reactants that react and products produced in solution-phase reactions
- Learn the RICE table problem-solving strategy for limiting reactant problems

Success Criterion

- Accurately calculate the amount of material that reacts and is produced in a chemical reaction in solution

Prerequisites

03-3 *Moles and Molar Mass*

04-1 *Balanced Chemical Reaction Equations*

04-2 *Limiting Reactants*

04-3 *Solution Concentration and Dilution*

MODEL 1 SOLUTION STOICHIOMETRY CALCULATIONS

Problem: Calculate the mass of $Ag_2CO_3(s)$ produced by combining 125 mL of 0.315 M $Na_2CO_3(aq)$ with 75.0 mL of 0.155 M $AgNO_3(aq)$. How many moles of excess reactant remain in solution?

Solution: As always, we need to start with a balanced chemical reaction equation:

$$2AgNO_3(aq) + Na_2CO_3(aq) \rightarrow Ag_2(CO_3)(s) + 2NaNO_3(aq)$$

Since the concentration and volume of both reactants are provided, we can determine the number of moles of each. Moreover, this is a limiting reactant problem, so it will be necessary to find the amount of product based on the limiting reactant.

Case I

If *silver nitrate* is the limiting reactant, here is the amount of silver carbonate produced:

$$\left(0.0750 \text{ L AgNO}_3\right)\left(\frac{0.155 \text{ mol AgNO}_3}{1 \text{ L AgNO}_3}\right)\left(\frac{1 \text{ mol Ag}_2(CO_3)}{2 \text{ mol AgNO}_3}\right)\left(\frac{275.75 \text{ g Ag}_2(CO_3)}{1 \text{ mol Ag}_2(CO_3)}\right)$$

$$= 1.60 \text{ g Ag}_2(CO_3)$$

Case II

If *sodium carbonate* is the limiting reactant, here is the amount of silver carbonate produced:

$$\left(0.125 \text{ L Na}_2CO_3\right)\left(\frac{0.315 \text{ Na}_2CO_3}{1 \text{ L Na}_2CO_3}\right)\left(\frac{1 \text{ mol Ag}_2(CO_3)}{1 \text{ mol Na}_2CO_3}\right)\left(\frac{275.75 \text{ g Ag}_2(CO_3)}{1 \text{ mol Ag}_2(CO_3)}\right)$$

$$= 10.9 \text{ g Ag}_2(CO_3)$$

Since the *smaller* amount of product identifies the limiting reactant, silver nitrate is the limiting reactant, and 1.60 g of silver carbonate is produced as the precipitate. By definition, the amount of the limiting reactant (silver nitrate) left over is zero.

The amount of sodium carbonate left over is found by subtracting the amount of it used in the reaction from the initial amount: (initial amount) – (amount used) = amount left over

Initial amount	Amount used
$= \left(0.125 \text{ L Na}_2\text{CO}_3\right)\left(\dfrac{0.315 \text{ Na}_2\text{CO}_3}{1 \text{ L Na}_2\text{CO}_3}\right)$	$= \left(0.0750 \text{ L AgNO}_3\right)\left(\dfrac{0.155 \text{ mol AgNO}_3}{1 \text{ L AgNO}_3}\right)\left(\dfrac{1 \text{ mol Na}_2\text{CO}_3}{2 \text{ mol AgNO}_3}\right)$
$= 0.039375$ moles Na_2CO_3	$= 0.0058125$ moles Na_2CO_3

$$\left(0.039375 \text{ moles Na}_2\text{CO}_3\right) - \left(0.0058125 \text{ moles Na}_2\text{CO}_3\right) = 0.0336 \text{ moles Na}_2\text{CO}_3$$

KEY QUESTIONS

1. How are the moles of each reactant found?

2. Why is this a limiting reactant problem?

3. How does this method determine the limiting reactant and amounts of products?

4. How does this method determine the amount of excess reactant remaining?

MODEL 2	RICE TABLE PROBLEM-SOLVING STRATEGY

Calculate the mass of $Ag_2CO_3(s)$ produced by combining 125 mL of 0.315 M $Na_2CO_3(aq)$ with 75.0 mL of 0.155 M $AgNO_3(aq)$ and the number of moles of the excess reactant remaining in solution.

Your calculations will be more efficient if you realize that 1 M = 1mmol/mL and that molar masses can be expressed in mg/mmol as well as g/mol (where 1mmol = 0.001 mol).

The RICE Table Solution

Steps in the Strategy	Example
Step R: Write the **Reaction equation**.	Both Na_2CO_3 and $AgNO_3$ dissociate in aqueous solution. The net ionic equation for the precipitation is: $$2\,Ag^+(aq) \;+\; CO_3^{2-}(aq) \;\rightarrow\; Ag_2CO_3(s)$$
Step I: Write the **Initial amounts** of reactants and products.	$\begin{array}{ccc} 11.6 & 39.4 & 0 \\ \text{mmol} & \text{mmol} & \end{array}$ $(75\text{ mL})(0.155\text{ M}) = 11.6\text{ mmol }Ag^+$ $(125\text{ mL})(0.315\text{ M}) = 39.4\text{ mmol }CO_3^{2-}$
Step C: Write the **Change** in the amounts due to the reaction, using the stoichiometric coefficients in the reaction equation.	In the reaction equation, the stoichiometric coefficient for silver is 2, the coefficient for carbonate is 1, and the coefficient for silver carbonate is 1. The equation is telling us that 2 moles of Ag^+ react with 1 mole of CO_3^{2-} to produce 1 mole of Ag_2CO_3. To get some amount of precipitate, say x moles, it will take x moles of carbonate and $2x$ moles of silver, because the amounts must be in the same ratio as the coefficients ($2:1:1 = 2x{:}x{:}x$). $\begin{array}{ccc} 2\,Ag^+(aq) \;+ & CO_3^{2-}(aq) \;\rightarrow & Ag_2CO_3(s) \\ 2x & x & x \end{array}$
Step E: Write the amounts of substances present after the reaction reaches **Equilibrium** or is complete.	For the reactants, subtract the amounts that react from the initial amounts present. For the products, add the amounts produced to the initial amount present. $\begin{array}{ccc} 2\,Ag^+(aq) \;+ & CO_3^{2-}(aq) \;\rightarrow & Ag_2CO_3(s) \\ 11.6 - 2x & 39.4 - x & 0 + x \end{array}$
Then solve for unknowns as requested by the problem statement.	The final amount of the limiting reactant will be 0. The limiting reactant is silver, so $$11.6 - 2x = 0 \text{ and therefore } x = 5.8 \text{ mmol.}$$ Now find how much carbonate remains and how much silver carbonate is produced. $$39.4 - 5.8 = \textbf{33.6 mmol } CO_3^{2-} \text{ remain}$$ $$(5.8\text{ mmol})(275.8\text{ g/mol}) = 1.60\text{ g }Ag_2CO_3 \text{ produced.}$$

Key Questions

5. How are the amounts of reactants initially present calculated?

6. What does the abbreviation *mmol* represent?

7. Why is Ag^+ and not CO_3^{2-} the limiting reactant?

8. Why is writing the balanced reaction equation an important part of the methodology?

9. Can you improve the methodology by changing the steps, changing the order of steps, or omitting some steps? Explain.

10. What insights about solving solution stoichiometry problems did your group gain from **Model 1** and the Key Questions?

1. Calculate the mass of $Fe(OH)_3(s)$ produced by mixing 30.0 mL of 0.138 M KOH(aq) and 15.0 mL of 0.196 M $Fe(NO_3)_3$(aq), and the number of moles of the excess reactant remaining in solution. The balanced reaction equation is:

$$Fe(NO_3)_3 \text{ (aq)} + 3KOH \text{ (aq)} \rightarrow Fe(OH)_3 \text{ (s)} + 3KNO_3 \text{ (aq)}$$

RICE Table Strategy Steps	Fill in the Details		
R: Write the reaction equation.			
I: Write the initial amounts of reactants and products.			
C: Write the change in amounts due to the reaction.			
E: Write the amounts after the reaction reaches equilibrium.			
Solve:			

2. Identify the excess reactant (HCl or NaOH), if any, and its final molar concentration when 1.75 g of NaOH(s) is stirred into 200.0 mL of 0.350 M HCl(aq). The balanced reaction equation is:

$$HCl(aq) + NaOH(aq) \rightarrow H_2O(l) + NaCl(aq)$$

RICE Table Strategy Steps	Fill in the Details
R: Write the reaction equation.	
I: Write the initial amounts of reactants and products.	
C: Write the change in amounts due to the reaction.	
E: Write the amounts after the reaction reaches equilibrium.	
Solve:	

1. One method of commercially removing the skins from potatoes is to soak them in a 3 to 6 M solution of sodium hydroxide for a short time. When the potatoes are removed from this solution, the skins are sprayed off. In one titration analysis of the NaOH solution, 36.2 mL of 0.650 M sulfuric acid (a diprotic acid) was required to neutralize (react completely) with a 25.0 mL sample of the soaking solution. Calculate the concentration of the NaOH soaking solution. Was it in the required range of 3 to 6 M?

The balanced reaction equation is: $H_2SO_4(aq) + 2NaOH(aq) \rightarrow 2H_2O(l) + Na_2SO_4(aq)$

RICE Table Strategy Steps	Fill in the Details
R: Write the reaction equation.	
I: Write the initial amounts of reactants and products.	
C: Write the change in amounts due to the reaction.	
E: Write the amounts after the reaction reaches equilibrium.	
Solve:	

Why?

Some ionic compounds dissolve and dissociate into ions when added to water. Other compounds are not soluble in water; these compounds will not dissolve, or may form a precipitate when solutions of soluble compounds are mixed. Compounds dissolving in water and precipitating from aqueous solutions are associated with health issues, environmental issues, and manufacturing processes. In order to utilize or evaluate the effects of a chemical reaction, a chemist (you!) must be able to identify its type and conditions under which it can occur, and describe it by a balanced reaction equation.

Learning Objective

- Recognize the characteristics of dissociation and precipitation of ionic compounds

Success Criteria

- Write and balance total ionic, net ionic, and molecular reaction equations

- Identify whether a precipitation reaction will occur when solutions are mixed

Prerequisites

03-1   *Nomenclature: Naming Compounds*

04-1   *Balanced Chemical Reaction Equations*

MODEL 1   WRITING DISSOCIATION AND PRECIPITATION REACTIONS OF IONIC COMPOUNDS

Many compounds dissolve and dissociate into ions when added to water, e.g.,

$$Na_2S(s) \rightarrow 2Na^+(aq) + S^{2-}(aq)$$

$$Cd(NO_3)_2(s) \rightarrow Cd^{2+}(aq) + 2NO_3^-(aq)$$

$$NaNO_3(s) \rightarrow Na^+(aq) + NO_3^-(aq)$$

Other compounds are insoluble in water, e.g., cadmium sulfide. So when solutions of sodium sulfide and cadmium nitrate are mixed together, solid cadmium sulfide forms. This solid is called a *precipitate*.

$$2Na^+(aq) + S^{2-}(aq) + Cd^{2+}(aq) + 2NO_3^-(aq) \rightarrow CdS(s) + 2Na^+(aq) + 2NO_3^-(aq)$$

This precipitation reaction equation is called a *total ionic reaction equation* because it includes all the ions in the solutions that were mixed together. Notice that the complete formula of the solid is written since it is not dissolved and dissociated into ions.

A precipitation reaction can also be written as a *net ionic equation*, where only the ions that participate in the precipitation reaction are included. The ions that don't participate are called *spectator ions*.

$$Cd^{2+}(aq) + S^{2-}(aq) \rightarrow CdS(s)$$

A reaction can also be written as a balanced molecular equation, where the (aq) designation means the species as it exists in aqueous solution, which in this case would be the ionic dissociation products. Keep in mind that even though the term *molecular equation* is used, these are ionic compounds, so a better term would be *formula equation*.

$$Na_2S(aq) + Cd(NO_3)_2(aq) \rightarrow CdS(s) + 2NaNO_3(aq)$$

KEY QUESTIONS

1. What is the characteristic identifying feature of a dissociation reaction?

2. How are the values of the balancing coefficients of ions produced in an ionic dissociation reaction related to the subscripts in the formula of the reactant ionic compound?

3. What is the characteristic identifying feature of a precipitation reaction?

4. What is common to the ionic reaction equation, the net ionic equation, and the molecular equation?

5. What is the characteristic identifying feature of each of the following: the ionic reaction equation, the net ionic equation, and the molecular equation?

INFORMATION

Ionic precipitation reactions are also called *double replacement reactions* or *metathesis reactions*. **Model 1** illustrated how soluble ionic compounds dissociate into separate ions in aqueous solution. When ions are dissociated, they are free to react independently of their counterions, so the total ionic equation emphasizes this separation as well as the number of ions of each type present after dissociation. In predicting whether a combination of solutions of ions will produce a precipitate we can use *solubility rules*. A list of rules is provided on the following page.

SOLUBILITY RULES FOR IONIC COMPOUNDS IN WATER

Soluble Ionic Compounds	Insoluble Exceptions
All common compounds of Group 1A (1) ions	None
All common compounds of ammonium ion	None
All common nitrates, acetates, and perchlorates	None
All common chlorides, bromides, and iodides	Ag^+, Pb^{2+}, Cu^+, and Hg_2^{2+}
All common fluorides	Pb^{2+} and Group 2A(2) cations
All common sulfates	Ca^{2+}, Sr^{2+}, Ba^{2+}, Ag^+, Pb^{2+}

Insoluble Compounds	Soluble Exceptions
All common metal hydroxides	Group 1A(1) cations and Ca^{2+}, Sr^{2+} and Ba^{2+}
All common carbonates and phosphates	Group 1A(1) cations and NH_4^+
All common sulfides	Group 1A(1) and Group 2A(2) cations and NH_4^+

MODEL 2   **PREDICTING PRECIPITATION REACTIONS OF IONIC COMPOUNDS**

Model 1 illustrated ionic dissociation reactions of soluble ionic compounds and how combining solutions of soluble ionic compounds can result in the formation of an insoluble precipitate. Since the dissociated ions act independently of each other in solution, they are free to react to form new combinations, as shown by the total ionic equation. Solubility rules allow one to predict the formation of a precipitate.

Consider the possible precipitants when mixing solutions of soluble silver nitrate and soluble sodium chloride. The reactants in the molecular equation are shown:

$$AgNO_3(aq) + NaCl(aq) \longrightarrow$$

The new possibilities for the product ionic compounds are determined by exchanging the ions, so Ag^+ can recombine with Cl^-, and Na^+ can recombine with NO_3^- giving the possible products as $AgCl$ and $NaNO_3$. Before completing the reaction equation, it is necessary to determine if either of the new combinations is insoluble according to the solubility rules. If either possible product is insoluble, then the reaction will form the precipitate.

According to the solubility rules, all common chloride salts are soluble, but $AgCl$ is an exception, so $AgCl$ will form a solid precipitate. Sodium salts fall under the category of soluble Group 1A compounds, so $NaNO_3$ remains in solution. The overall reaction is then:

$$AgNO_3(aq) + NaCl(aq) \longrightarrow AgCl(s) + NaNO_3(aq)$$

Of course, if there is no insoluble combination formed, there will not be a precipitation reaction. It is customary to write ***N.R.*** for "no reaction" after the arrow instead of products in these cases.

6. Why are new combinations of cations and anions easily formed in mixing solutions of soluble ionic compounds?

7. Why is it necessary to remember the names, formulas, and ionic charges of polyatomic ions in working with precipitation reactions?

8. Write the net ionic equation for the precipitation reaction in **Model 2**.

EXERCISES

1. For each of the pairs of reactants in a) through c) that are shown on the next page, do the following:

   i. Write the formulas for the reactants.

   ii. Below these formulas, write the ions that these reactants produce through dissociation in aqueous solution.

   iii. Predict whether a precipitation reaction will occur. Consider the possible compounds produced when the anions and cations of the reactants exchange partners. If any combination is insoluble according to solubility rules,, then the reaction is a precipitation reaction.

   iv. Write the balanced overall ionic reaction using the formulas for the reactants and products. Be sure to include the states, in parentheses, for each.

   v. Write the balanced net ionic equation, omitting the spectator ions.

   vi. Write the balanced molecular equation.

a. lead(II) acetate and sodium carbonate in water

    i. _____

    ii. _____

    iii. _____

    iv. _____

    v. _____

    vi. _____

b. calcium chloride and sodium phosphate in water

    i. _____

    ii. _____

    iii. _____

    iv. _____

    v. _____

    vi. _____

c. ammonium chloride and copper(II) sulfate in water.

    i. _____

    ii. _____

    iii. _____

    iv. _____

    v. _____

    vi. _____

05-2 Introduction to Acid – Base Reactions

WHY?

Aqueous reactions involving the transfer of protons from acids to bases are very important in all disciplines of chemistry and their applications. The terms *acid* and *base* are actually used for several classes of reactions, but the emphasis on the Brønsted-Lowry definition is most important for general chemistry.

LEARNING OBJECTIVE

- Identify the characteristics of acids and bases

SUCCESS CRITERIA

- Quickly recognize acids and bases
- Write acid-base neutralization reactions correctly

PREREQUISITES

03-1 *Nomenclature: Naming Compounds*

04-1 *Balanced Chemical Reaction Equations*

INFORMATION

According to the Brønsted-Lowry definition of acids and bases:

An *acid* is a chemical species that donates a proton to another species.

A *base* is a chemical species that accepts a proton from another species.

According to the Arrhenius definition of acids and bases:

Acids undergo ionization in water to form the hydronium ion, which can be written as $H^+(aq)$ or $H_3O^+(aq)$. In the presence of an acid, water acts as a Brønsted-Lowry base.

Bases include soluble hydroxides that generate $OH^-(aq)$ in solution. Hydroxide ion can also be generated by deprotonation of water by a Brønsted-Lowry base. In the presence of a base, water acts as a Brønsted-Lowry acid.

MODEL 1   RECOGNIZING ACIDS AND BASES

Acids were briefly introduced previously, in the study of nomenclature, and some common binary acids and oxo-acids were given as examples. In all of these cases, the formulas indicated the ionizable protons as the first element written in the formula, as in HCl and HNO_3.

Of these acids, there are six that you should remember as *strong acids*, which undergo virtually 100% complete ionization in water. They are:

$$HCl, HBr, HI, HNO_3, HClO_4 \text{ and } H_2SO_4$$

While there are other examples of strong acids, focus on these common ones for now. The rest of the binary and oxo acids are *weak acids*; that is, they undergo ionization to a small extent and establish an *equilibrium*. Among these acids are:

$$HF, HNO_2, H_3PO_4, \text{ etc.}$$

Another important category of acids is *organic* or *carboxylic acids*, that contain the –COOH functional group. Acetic acid, which is found in vinegar, is a carboxylic acid with the formula CH_3COOH. Notice that the proton on the –COOH group is the acidic one and is donated on ionization. Carboxylic acids are generally weak.

Strong bases are soluble hydroxides, since they produce 100% of the available hydroxide ion concentration as they dissolve. Most hydroxides are insoluble, and may be considered *weak bases*.

Weak bases include ammonia, NH_3 and organic amines, which resemble ammonia, with the H atom replaced by a carbon-containing group, such as the $–CH_3–$ methyl group in methyl amine, CH_3NH_2.

KEY QUESTION

1. Indicate the acid or base reactivity of the following species with water and whether the reaction goes to completion or not.

   a. HBr

   b. H_2SO_3

   c. KOH

   d. $Mg(OH)_2$

   e. H_2CO_3

   f. HCOOH

EXERCISE

1. Label each of the reactants in **Model 2**, which follows, as *acid* or *base*, using the definitions of an acid and a base given previously.

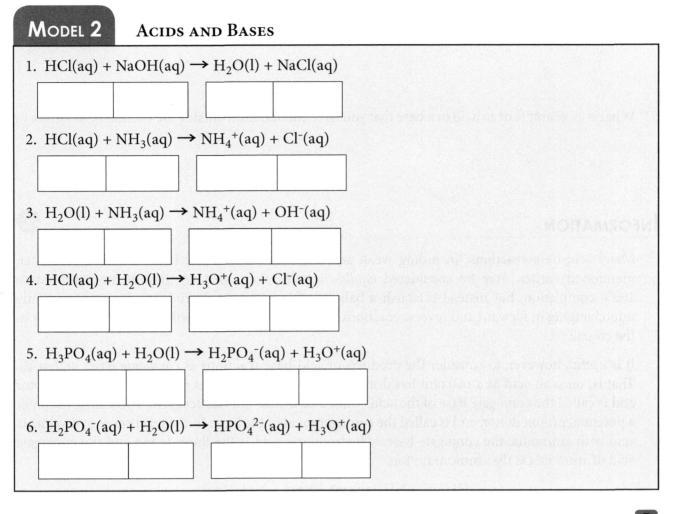

1. $HCl(aq) + NaOH(aq) \rightarrow H_2O(l) + NaCl(aq)$

2. $HCl(aq) + NH_3(aq) \rightarrow NH_4^+(aq) + Cl^-(aq)$

3. $H_2O(l) + NH_3(aq) \rightarrow NH_4^+(aq) + OH^-(aq)$

4. $HCl(aq) + H_2O(l) \rightarrow H_3O^+(aq) + Cl^-(aq)$

5. $H_3PO_4(aq) + H_2O(l) \rightarrow H_2PO_4^-(aq) + H_3O^+(aq)$

6. $H_2PO_4^-(aq) + H_2O(l) \rightarrow HPO_4^{2-}(aq) + H_3O^+(aq)$

KEY QUESTIONS

2. For the reactions in **Model 2,** how did you determine whether a reactant was an acid or a base?

3. Does any substance in the model act both a proton donor (acid) and as a proton acceptor (base)? Explain how this can or cannot occur. A substance that can act either as an acid or a base is called amphoteric; an amphoteric substance that can either donate or accept a proton is also called amphiprotic.

4. Would you call sodium hydroxide, NaOH, an acid or a base? Why? Sodium hydroxide ionizes in water to give $Na^+(aq)$ and $OH^-(aq)$.

5. What is an example of an acid or a base that you have encountered outside the chemistry laboratory?

INFORMATION

Many acid-base reactions, including weak acid ionization and weak base ionization in water, mentioned earlier, may be considered *equilibrium reactions*; that is, they do not proceed to 100% completion, but instead establish a balance of reactants and products that are constantly interchanging in forward and reverse reactions. Equilibrium reactions will be a major topic later in the course.

It is useful, however, to consider the products of acid-base reactions as *conjugate bases and acids*. That is, once an acid as a reactant has donated its proton, it becomes a potential proton acceptor and is called the conjugate base of the acid. Once a base reactant has accepted a proton, it becomes a potential proton donor, and is called the conjugate acid of the base. In the reaction of hydrofluoric acid with ammonia, the conjugate base of hydrofluoric acid is the fluoride ion and the conjugate acid of ammonia is the ammonium ion.

$$HF(aq) + NH_3(aq) \rightarrow F^+(aq) + NH_4^+(aq)$$

EXERCISES

2. Label the acids on the right-hand side of the reaction equations in **Model 2**.

3. Label the bases on the right-hand side of the reaction equations in **Model 2**.

4. Write the acid-base reaction for NH_3 reacting with HNO_2, and identify the acid and the base on both sides.

5. Complete the following table for the reaction equations in **Model 2**. Label each reactant and product as *acid* or *base*. Reaction 1 is done for you.

Reaction	Reactant	Corresponding Product
1	HCl (acid)	Cl^- (base)
1	NaOH (base)	H2O (acid)

Reaction	Reactant	Corresponding Product
2		
2		
3		
3		
4		
4		
5		
5		
6		
6		

6. Examine your entries in the table in Exercise 5 and identify how the acids and the bases in each row differ.

7. Complete the following table of conjugate acid-base pairs.

	Acid	Base
a.		CN^-
b.	HNO_3	
c.	H_2CO_3 (a diprotic acid)	
d.		CO_3^{2-}

	Acid	Base
e.		NH_2^-
f.	$HClO_4$	
g.		CH_3NH_2
h.	$(CH_3)_2NH_2^+$	

8. For the following reaction, identify the reactant that is an acid, the reactant that is a base, and the two conjugate acid-base pairs present.

$$H_2SO_4 + HPO_4^{2-} \rightarrow HSO_4^- + H_2PO_4^-$$

WHY?

Corrosion of metals, combustion of fuels, the generation of electricity from batteries, and many biological processes involve electron transfer reactions. In these reactions, which are also called oxidation-reduction or redox reactions, electrons are transferred from one chemical species to another. By understanding the principles of redox reactions, scientists and engineers can prevent corrosion, design conditions for more efficient combustion, produce new kinds of batteries, and increase the lifespan of materials and biological systems.

LEARNING OBJECTIVES

- Distinguish electron transfer reactions from other kinds of reactions
- Characterize electron transfer reactions in terms of electron flow and changes in oxidation numbers
- Efficiently balance oxidation-reduction reactions in acidic and basic conditions

SUCCESS CRITERIA

- Assign oxidation numbers to atoms in a chemical compound
- Identify the oxidizing agent and the reducing agent in a redox reaction
- Identify the number of electrons transferred in a redox reaction
- Systematically balance redox reactions

PREREQUISITES

02-3   *The Periodic Table of the Elements*

03-1   *Nomenclature: Naming Compounds*

04-1   *Balanced Chemical Reaction Equations*

INFORMATION

The *oxidation number* of an atom is the charge an atom would have if it were part of an ionic bond in which all valence electrons were transferred to the less metallic atom. Oxidation numbers help you keep track of electron flow in redox reactions, even though electrons may not be transferred completely. The oxidation number of an atom in a chemical compound is the basic tool used to identify and understand redox reactions. The oxidation number is defined as zero for the elemental state, such as sodium metal or molecular oxygen.

When an atom is *oxidized*, it loses electrons, and its oxidation number increases. When an atom is *reduced*, it gains electrons, and its oxidation number decreases. The is most clearly seen in monatomic ions, such as Na^+ or Cl^-, in which the ionic charge is equal to the oxidation number.

In molecules or polyatomic ions, each atom is assigned an oxidation number. Some elements have very predictable oxidation numbers and these will be discussed in the model and used as a starting point to calculate the oxidation number of an unknown variable. In all cases, the sum of the oxidation numbers in an ion is equal to its ionic charge.

The reactant containing the element being oxidized is called the *reducing agent,* because it causes another reactant to be reduced by providing it with electrons. The reactant containing the atom being reduced is called the *oxidizing agent,* because it causes the other reactant to be oxidized by accepting electrons.

To summarize: *Oxidation* is the loss of electrons from a chemical species. *Reduction* is the gain of electrons by a chemical species. To remember these definitions, think: *reduction* means the *oxidation number* is *reduced.* Another common mnemonic device is the acronym **OIL-e-RIG**, which stands for **O**xidation **I**s **L**oss of **e**lectrons, **R**eduction **I**s **G**ain.

There are two other ways to quickly recognize oxidation and reduction reactions. Gain of oxygen atoms is generally associated with oxidation, while loss of oxygen atoms is associated with reduction. Likewise, gain of hydrogen atoms (not protons as in acid-base reactions) is associated with reduction, while loss of hydrogen atoms is associated with oxidation.

MODEL 1 GUIDELINES FOR ASSIGNING OXIDATION NUMBERS

The key to understanding oxidation numbers is to remember that the *oxidation number* is defined as the charge an atom would have if each of its valence (or bonding) electrons were assigned to the less metallic atom in each of its bonds.

Chemical Species	Oxidation Number	Rationale
O_2	0	The oxidation number of an atom of a pure element is always 0.
Fe metal	0	
He	0	The oxidation number of an atom or a monatomic ion is its charge.
Mg^{2+}	+2	
Cl^-	−1	
H_2O	H = +1	Hydrogen in a compound has an oxidation number of +1, unless it is combined with a metal such as sodium or calcium, in which case it is −1.
CH_4		
LiH	H = −1	
NaH		
HF	F = −1	Fluorine in a compound always has an oxidation number of −1.
CF_4	F = −1	
H_2O	O = −2	Oxygen in a compound has an oxidation number of −2 unless it is a peroxide, in which case it is −1. If oxygen is combined with fluorine, the oxygen isn't the most electronegative element (fluorine is) and has an oxidation state of +2.
H_2O_2	O = −1	
Na_2O_2		

Chemical Species	Oxidation Number	Rationale
OF_2	$+2(O) - 1(F) \times 2 = 0$	*In a polyatomic molecule or ion, the sum of all the oxidation numbers of the atoms is equal to the charge on the molecule or ion.*
BrCl	Cl = −1 Br = +1	*If the previous rules do not determine all the oxidation numbers, the less metallic atom is assigned an oxidation number equal to the charge it would have if it had 8 valence electrons.*

KEY QUESTIONS

1. What is the oxidation number of a pure element such as H_2, N_2, K, Ni, S_8, or Cu?

2. For the polyatomic ion SO_4^{2-}, what is the sum of the oxidation numbers of sulfur and the four oxygen atoms?

3. Without calculating the oxidation numbers of nitrogen, predict which species would have a more positive oxidation number. Circle that species. Then justify your prediction. Afterwards, calculate the oxidation numbers to verify your prediction.

   a. NH_3 or N_2

   b. NO or N_2

   c. NO_2^- or NO_3^-

GOT IT!

Explain why you agree or disagree with each of the following statements.

1. The least metallic atom in a compound will generally have a negative oxidation number.

2. The most metallic atom in a compound will have a positive oxidation number.

3. The oxidation number of the manganese atom in the permanganate ion is –1.

4. The sum of all oxidation numbers must equal the charge on the species. For a neutral compound the sum must be 0; for a cation like NH_4^+ the sum must be +1; and for an anion like SO_4^{2-}, the sum must be –2.

EXERCISES

1. Using the guidelines in the model and the insight gained from your answers to the *Key Questions* and *Got It!* sections, assign oxidation numbers to all atoms in the following compounds and ions.

O_2	H_2O	CH_4	CO_2	SF_6
___	___	___	___	___
$NaNO_3$	SO_4^{2-}	H_2O_2	$CuCl_2$	$HClO_2$
___	___	___	___	___
$KMnO_4$	HCN	H_2S	P_4	Na_2SO_3
___	___	___	___	___
	K_2CrO_4	Cr_2O_3	$Na_2Cr_2O_7$	
	___	___	___	

MODEL 2   **ELECTRON TRANSFER REACTIONS**

In the oxidation of copper metal, electrons are transferred from copper atoms to molecular oxygen to form copper oxide. The chemical reaction equation is

$$2Cu + O_2 \rightarrow 2CuO$$

The oxidation number of each copper atom changes from 0 to +2 because two electrons were lost, and the oxidation number of each oxygen atom changes from 0 to −2 because two electrons were gained.

Oxygen is reduced because the oxidation number decreases, and copper is oxidized because the oxidation number increases. Oxygen is therefore the oxidizing agent, and copper is the reducing agent. You can recognize a redox reaction by determining whether oxidation numbers change or not.

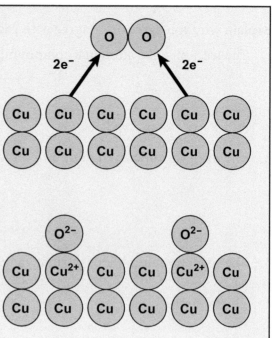

KEY QUESTIONS

4. In the oxidation of copper, how many electrons are transferred from one copper atom to one oxygen atom?

5. What are the oxidation numbers of the atoms in Cu metal, O_2, and CuO?

6. In the oxidation of copper, what is the chemical species that is oxidized, and what is the chemical species that is reduced?

7. In the oxidation of copper, what is the oxidizing agent, and what is the reducing agent?

8. If the oxidation number of a chemical species increases during a chemical reaction, does that species act as the oxidizing agent or the reducing agent?

9. How can you distinguish a redox reaction from other types of chemical reactions?

2. Assign oxidation numbers to all the atoms in the following reactions. For the redox reactions, identify the atom oxidized, the atom reduced, the oxidizing agent, and the reducing agent.

a.
$$CO_2(g) + H_2(g) \rightarrow 2\,CO(g) + H_2O(g)$$

Redox (yes/no)	Atom Oxidized	Atom Reduced	Oxidizing Agent	Reducing Agent

b. This reaction is used to produce industrially important quicklime from limestone.

$$CaCO_3(s) \rightarrow CaO(s) + CO_2$$

Redox (yes/no)	Atom Oxidized	Atom Reduced	Oxidizing Agent	Reducing Agent

c. This reaction can be used to produce hydrogen.

$$2H_3O^+(aq) + Mg(s) \rightarrow Mg^{2+}(aq) + H_2(g) + 2H_2O(aq)$$

Redox (yes/no)	Atom Oxidized	Atom Reduced	Oxidizing Agent	Reducing Agent

d.
$$Cl_2(g) + NaBr(s) \rightarrow Br_2(g) + NaCl(s)$$

Redox (yes/no)	Atom Oxidized	Atom Reduced	Oxidizing Agent	Reducing Agent

e. This reaction occurs in an automobile battery.

$$Pb(s) + 2HSO_4^-(aq) + PbO_2(s) + 2H_3O^+(aq) \rightarrow 2PbSO_4(s) + 4H_2O(l)$$

Redox (yes/no)	Atom Oxidized	Atom Reduced	Oxidizing Agent	Reducing Agent

f.
$$HF(aq) + NaOH(aq) \rightarrow NaF(aq) + H_2O(aq)$$

Redox (yes/no)	Atom Oxidized	Atom Reduced	Oxidizing Agent	Reducing Agent

MODEL 3   BALANCING SIMPLE ELECTRON TRANSFER REACTIONS

Redox reactions involving metals, metal ions, hydrogen, and the hydronium ion are relatively easy to balance since the only consideration when determining the oxidation numbers is the ionic charge. Even in these cases, however, it is helpful and instructive to separate the overall redox reaction into half reactions: the oxidation reaction shows the lost electrons as products, while the reduction reaction shows the gained electrons as reactants. In the overall redox reaction, however, electrons lost must be equal to electrons gained, so that the two will cancel out when the half reactions are added.

The skeletal, unbalanced redox reaction shows the products and reactants with their appropriate ionic charges.

$$Ag^+(aq) + Cu(s) \rightarrow Ag(s) + Cu^{2+}(aq)$$

To balance this reaction, first separate the overall reaction into half reactions.

$$Ag^+(aq) \rightarrow Ag(s) \qquad\qquad Cu(s) \rightarrow Cu^{2+}(aq)$$

Then determine the total ionic charge for the products and the reactants for each half reaction.

$$Ag^+(aq) \rightarrow Ag(s) \qquad\qquad Cu(s) \rightarrow Cu^{2+}(aq)$$
$$+1 \qquad\quad 0 \qquad\qquad\qquad 0 \qquad\quad +2$$

Then, add negatively charged electrons, written as e$^-$, to the reactants or products in order to balance the ionic charge.

$$e^- + Ag^+(aq) \rightarrow Ag(s) \qquad\qquad Cu(s) \rightarrow Cu^{2+}(aq) + 2e^-$$

When adding the two half reactions together, the number of electrons must cancel. In order for that to happen, the silver half reaction must be multiplied by a factor of two, so that its two electrons lost cancel out the two electrons gained by copper.

$$2e^- + 2Ag^+(aq) \rightarrow 2Ag(s)$$

$$+ \qquad Cu(s) \rightarrow Cu^{2+}(aq) + 2e^-$$

$$\overline{\qquad 2Ag^+(aq) + Cu(s) \rightarrow 2Ag(s) + Cu^{2+}(aq)}$$

KEY QUESTIONS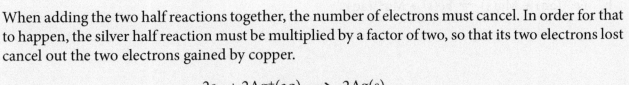

10. Based on the change in oxidation numbers, in **Model 3**, which metal is oxidized and which is reduced?

11. Is the need for electrons in the balanced half reaction equations consistent with your answer to the previous question? Explain.

12. Why is it necessary for the number of electrons shown in the half reactions to cancel as they are added together to give the overall redox reaction?

EXERCISES

3. Balance the following redox reactions using the half reaction method.

   a. $Zn(s) + H^+(aq) \rightarrow Zn^{2+}(aq) + H_2(g)$

b. $Fe^{3+}(aq) + Mg(s) \rightarrow Fe(s) + Mg^{2+}(aq)$

MODEL 4	BALANCING ELECTRON TRANSFER REACTIONS INVOLVING WATER UNDER ACIDIC AND BASIC CONDITIONS

Redox reactions taking place in aqueous solution can involve water, hydronium ions, or hydroxide ions as a source or sink for oxygen and hydrogen atoms. In these cases it is necessary to use a systematic method for balancing the overall reactions, again using the half reaction approach. The method used here is the ion-electron method, which focuses on the balance of ionic charge to determine the number of electrons lost and gained in the half reactions. A mnemonic will also be introduced to help you keep track of the steps in this method.

Consider the reaction of permanganate ion with iodide ion to produce manganese (II) and iodine in acidic conditions. The skeletal reaction is:

$$MnO_4^-(aq) + I^-(aq) \rightarrow Mn^{2+}(aq) + I_2(aq)$$

As with simple redox reactions, start by separating the overall reaction into half reactions:

Half Reactions $MnO_4^-(aq) \rightarrow Mn^{2+}(aq)$ $I^-(aq) \rightarrow I_2(aq)$

Since all the elements must be balanced, continue by balancing any element other than hydrogen and oxygen. Iodide needs a balancing coefficient of 2 to balance with I_2.

Balance Elements $MnO_4^-(aq) \rightarrow Mn^{2+}(aq)$ $2I^-(aq) \rightarrow I_2(aq)$

Since permanganate contains four oxygen atoms, they need to be accounted for in its half reaction. Water molecules are produced as permanganate loses oxygen, so add enough water molecules to the product side to balance the equation.

Oxygen Balance $MnO_4^-(aq) \rightarrow Mn^{2+}(aq) + 4H_2O(l)$ $2I^-(aq) \rightarrow I_2(aq)$

Now, since hydrogen atoms have been introduced, they must be balanced by adding hydronium ions to the opposite side.

Hydrogen Balance $8H^+(aq) + MnO_4^-(aq) \rightarrow Mn^{2+}(aq) + 4H_2O(l)$ $2I^-(aq) \rightarrow I_2(aq)$

Each half reaction is now balanced with respect to the elements, but not the ionic charge. As in the previous model, electrons must be added as reactants or products to balance the ionic charge.

$$8H^+(aq) + MnO_4^-(aq) \rightarrow Mn^{2+}(aq) + 4H_2O(l) \qquad 2I^-(aq) \rightarrow I_2(aq)$$

$$(8)(+1) - 1 = +7 \qquad +2 + (4)(0) = +2 \qquad (2)(-1) = -2 \qquad 0$$

Five electrons are needed as reactants in the manganese reduction reaction, while two electrons are needed as products in the iodine oxidation reaction.

Charge Balance $5e^- + 8H^+(aq) + MnO_4^-(aq) \rightarrow Mn^{2+}(aq) + 4H_2O(l)$ $2I^-(aq) \rightarrow I_2(aq) + 2e^-$

The mnemonic created by the underlined letters is "**HBO-Hy-Charge**", and results in balanced half reactions.

To complete the process, the half reactions must be added with cancelation of electrons. In this case, a factor of two for the manganese reduction reaction will give 10 electrons, while a factor of five for the iodine oxidation will also give 10 electrons.

$$10e^- + 16H^+(aq) + 2MnO_4^-(aq) \rightarrow 2Mn^{2+}(aq) + 8H_2O(l)$$

$+$ $\underline{10I^-(aq) \rightarrow 5I_2(aq) + 10e^-}$

$$16H^+(aq) + 2MnO_4^-(aq) + 10I^-(aq) \rightarrow 2Mn^{2+}(aq) + 8H_2O(l) + 5I_2(aq)$$

If the conditions for this reaction were basic, an adjustment would be made at this point to account for the 16 hydronium ions in the reactants. By adding an equal number of hydroxide ions as hydronium ions to both sides of the reaction equation, the hydronium ions are effectively neutralized to make water, leaving hydroxide ions on the products side.

$16H^+(aq) + 2MnO_4^-(aq) + 10I^-(aq) \rightarrow 2Mn^{2+}(aq) + 8H_2O(l) + 5I_2(aq)$

$+\ 16OH^-(aq)$ $+\ 16OH^-(aq)$

$\overline{16H_2O(l) + 2MnO_4^-(aq) + 10I^-(aq) \rightarrow 2Mn^{2+}(aq) + 8H_2O(l) + 5I_2(aq) + 16OH^-(aq)}$

Subtracting 8 water molecules from each side simplifies the reaction equation.

$$8H_2O(l) + 2MnO_4^-(aq) + 10I^-(aq) \rightarrow 2Mn^{2+}(aq) + 5I_2(aq) + 16OH^-(aq)$$

KEY QUESTIONS

13. How can you tell when you will need to use this more complicated method for balancing redox reactions?

14. How is this method similar to the one in **Model 3**? How is it different?

EXERCISES

4. Balance the following redox reactions using the ion electron method under acidic conditions.

$$Cr_2O_7^{2-}(aq) + NH_4^+(aq) \rightarrow Cr^{3+}(aq) + NO_3^-(aq)$$

05-4 Analytical Titrations

WHY?

Titration is a long-used and very common technique for analyzing the amounts of substances (called *analytes*) present in solution. Titration relies on the analyte's reactivity in solution with other substances (called *titrants*) which have a known concentration. Titrations are used widely in the analysis of many foods and commercial products, as well as environmental samples.

LEARNING OBJECTIVES

- Apply balanced solution phase reactions to a titration analysis
- Recognize the correct terms in dimensional analysis of titration experiments

SUCCESS CRITERIA

- Accurately calculate the amount of analyte in solution based on titration data
- Accurately calculate the volume of titrant required to reach the equivalence point

PREREQUISITES

03-1 *Nomenclature: Naming Compounds*

04-1 *Balanced Chemical Reaction Equations*

04-3 *Solution Concentration and Dilution*

04-4 *Solving Solution Stoichiometry Problems*

05-2 *Introduction to Acid – Base Reactions*

05-3 *Electron Transfer Reactions*

INFORMATION

Different reaction types, including acid-base neutralization and oxidation-reduction, can be the basis for a titration, but the reaction must be reasonably fast and quantitative, going essentially to completion. Of course, the correct reaction stoichiometry must also be known and used in calculations based on the titration data. It is also necessary to be able to determine precisely and accurately when the proper stoichiometric amounts of the reactant solutions have been combined, using a colored indicator or an instrumental method such as a pH meter. The theoretical volume needed to reach a stoichiometric balance is called the *equivalence point*. The experimental measurement of the equivalence point is called the *end-point*.

Titrations are generally carried out with a known volume of analyte solution, to which the titrant is added using a buret. This allows determination of the volume required to reach the end-point. These are depicted in Figure 1 for an acid-base titration.

Figure 1

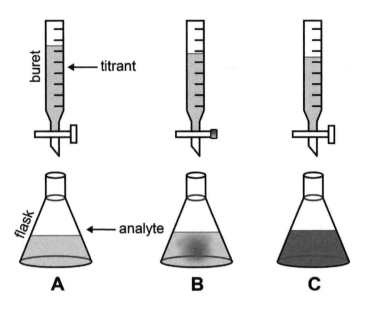

$$\text{NaOH(aq) (titrant)} + \text{HCl(aq) (analyte)} \rightarrow \text{H}_2\text{O(l)} + \text{NaCl(aq)}$$

MODEL 1   **TITRATION CALCULATIONS IN ACID-BASE NEUTRALIZATIONS**

Consider the acid-base titration in Figure 1 of the Information section between NaOH and HCl.

If the volume of HCl solution was originally 25.00 mL and the total volume of 0.2015 M NaOH solution required to reach the end-point was measured to be 35.06 mL, the concentration of HCl in the original solution is:

$$= \left(\frac{35.06 \ \text{mL NaOH}}{25.00 \ \text{mL HCl}} \right) \times \left(\frac{0.2015 \ \text{mmol NaOH}}{\text{mL NaOH}} \right) \times \left(\frac{1 \ \text{mmol HCl}}{1 \ \text{mmol NaOH}} \right)$$

$$= \frac{0.28258 \ \text{mmol HCl}}{\text{mL HCl}} = 0.2826 \ \text{M HCl}$$

If the titration involved the weak acid phosphoric acid, H_3PO_4, instead of HCl, the reaction molecular equation would be different, because phosphoric acid is a tri-protic acid.

$$3 \ \text{NaOH(aq) (titrant)} + \text{H}_3\text{PO}_4\text{(aq) (analyte)} \rightarrow 3 \ \text{H}_2\text{O(l)} + \text{Na}_3\text{PO}_4\text{(aq)}$$

Again, if the volume of acid was 25.00 mL and the total volume of 0.2015 M NaOH solution required to reach the end-point was measured to be 35.06 mL, the concentration of phosphoric acid in the original solution can be calculated as before:

$$= \left(\frac{35.06 \ \text{mL NaOH}}{25.00 \ \text{mL H}_3\text{PO}_4} \right) \times \left(\frac{0.2015 \ \text{mmol NaOH}}{\text{mL NaOH}} \right) \times \left(\frac{1 \ \text{mmol H}_3\text{PO}_4}{3 \ \text{mmol NaOH}} \right)$$

$$= \frac{0.094193 \ \text{mmol H}_3\text{PO}_4}{\text{mL H}_3\text{PO}_4} = 0.09419 \ \text{M H}_3\text{PO}_4$$

KEY QUESTIONS

1. How are the balancing coefficients in the two reactions in the model determined?

2. In calculations like the ones in the model, why is it acceptable to use the ratio of mL base/mL acid as the first term?

3. Why is it sensible to put the volume of base in the numerator and the volume of acid in the denominator in the first term of the calculation?

4. Why are *mmoles* used in the ratio of acid to base?

EXERCISES

1. In a titration of 10.00 mL of sulfuric acid of an unknown concentration, 28.62 mL of 0.1247 M sodium hydroxide was required to reach the endpoint. What was the concentration of the sulfuric acid?

2. A solution of ammonia, NH_3, can be titrated with a strong acid like HCl to yield ammonium chloride according to the reaction equation:

3. Oxalic acid, $H_2C_2O_4$, can be oxidized by potassium permanganate, $KMnO_4$, in acidic solution. The products of the reaction are $CO_2(g)$ and Mn^{2+} ion. What is the end-point volume of added potassium permanganate solution, in mL, if the oxalic acid concentration is 0.1506 M and its volume is 10.00 mL, and the potassium permanganate concentration is 0.2508 M?

WHY?

Chemical reactions release or store energy, usually in the form of *thermal energy*. Thermal energy is the kinetic energy of the atoms and molecules comprising an object. Heat is thermal energy that is transferred from one object to another. The amount of energy released or stored by a chemical reaction can be determined from the change in temperature. You need to know how much energy is either released or absorbed by a reaction in order to determine its utility. This knowledge also allows you to identify the conditions necessary for that reaction to occur efficiently and safely.

LEARNING OBJECTIVES

- Quantify the relationship between energy transferred and change in temperature
- Understand the utility of specific heat capacity
- Understand how to determine the heat of chemical reactions

SUCCESS CRITERIA

- Correctly use the specific heat capacity of substances in various situations to perform calculations relating heat and change in temperature
- Determine the energy transferred in a chemical reaction from the change in temperature

PREREQUISITES

04-1 *Balanced Chemical Reaction Equations*

05-2 *Introduction to Acid – Base Reactions*

INFORMATION

Energy is measured using SI units. One Joule (J) is defined as $1 \, kg \times m^2 \times s^{-2}$, while one calorie is defined as the amount of heat required to raise the temperature of 1g of water by 1°C.

The nutritional calorie which with you are familiar is equal to one kcal, or kilocalorie (1000 calories.)

Thermal energy is transferred spontaneously from an object at a higher temperature to another object at a lower temperature. Thermal equilibrium is reached when the two objects are at the same temperature.

The temperature of an object increases when it absorbs energy; the temperature of an object decreases when it releases energy. If specific heat is known, this temperature change can be used to determine the amount of energy that was absorbed or released.

MODEL 1   MEASURING THERMAL ENERGY

In an experiment to determine the relationship between the temperature change of water and the amount of heat it absorbs, an electrical heater was used to increase the temperature of 1 g of water by 1 °C. The heat produced by an electrical heater over some period of time can be calculated very accurately: heat = electrical current × voltage × time. The results of this experiment are given below.

Mass of Water	Change in Temperature	Energy Used in Heating the Water
1.000 g	1.000 °C	4.184 J

KEY QUESTIONS

1. How much energy was required to increase the temperature of 1.000 g of water by 1.000 °C?

2. How much energy would be required to increase the temperature of 100.0 g of water by 1.000 °C?

3. How much energy would be required to increase the temperature of 1.000 g of water by 25.0 °C?

4. The specific heat capacity of any substance is defined as the amount of energy needed to increase the temperature of 1 g of the substance by 1 °C. What is the specific heat capacity of water in units of J/g °C. Based on your answer and the information section, what is the relationship between Joules and calories?

5. Based on your answers to Key Questions 1-4, what is the equation that relates the amount of energy absorbed or released, by any substance, to the change in temperature of that substance? Write this equation using the following notation: q = heat in J, m = mass in g, ΔT = change in temperature in °C or Kelvin, since the increment is the same for both units, and c_s = specific heat capacity of the substance in J/g °C. Note that the Δ (Greek letter delta) sign indicates the change operator as (final state) – (initial state), so, $\Delta T = T_f - T_i$.

6. If you needed to experimentally determine the specific heat capacity of a new metal alloy that you had developed, what would you need to measure, and how would you use the measured values to calculate specific heat capacity?

EXERCISES

1. Aluminum has a specific heat capacity of 0.902 J/g °C. How much energy is released when 1.5 kg of aluminum cools from 35 °C to 20 °C?

2. As part of your morning routine, you make black tea to have with your breakfast. You boil 300 g of water (ending temperature of 100 °C) using 100 kJ energy from the electric kettle. What is the starting temperature of the water?

Got It!

1. At room temperature, equal masses of water and aluminum were each heated with 1 kJ of electrical energy. The temperature of the aluminum increased much more than that of the water. Explain why.

MODEL 2	HEAT OF A CHEMICAL REACTION

Solutions of hydrochloric acid and sodium hydroxide are mixed in a beaker that is surrounded by Styrofoam insulation. The temperature of the solution and beaker increases from 24 °C to 38 °C.

The volume of the resulting solution is 435 mL. Determine the amount of energy, ΔE_{total}, released by this reaction.

$$\Delta E_{total} = \Delta E_{solution} + \Delta E_{beaker}$$

$\Delta E_{total} = $ total energy released

$\Delta E_{solution} = $ energy used to heat the solution

$\Delta E_{solution} = c_s\, m\, \Delta T$

$\Delta E_{beaker} = $ energy used to heat the beaker, insulation, and thermometer

$\Delta E_{beaker} = (330\ \text{J/°C})\, \Delta T$

Assume that since the solution is dilute, its specific heat is the same as that of water. The beaker, insulation, and thermometer have a combined heat capacity of 330 J/°C. The *heat capacity* is the amount of energy required to change the temperature of some object by 1°C.

$$\Delta E_{total} = (4.184\ \text{J/g °C})(435\ \text{mL} \times 1.00\ \text{g/mL})(14\ \text{°C}) + (330\ \text{J/°C})(14\ \text{°C}) = 30.1\ \text{kJ}$$

Key Questions

7. In **Model 2**, what things become hotter due to the energy that is released in the reaction of hydrochloric acid and sodium hydroxide? What is the value of ΔT?

---

8. How is the energy that goes into heating the solution calculated in **Model 2**?

9. How is the energy used in heating the other items calculated in **Model 2**?

10. How is the total energy released in the chemical reaction in **Model 2** calculated?

11. What is the difference between the specific heat capacity and the heat capacity of a substance or object? Why are the units different?

PROBLEMS

1. You mix 100 mL of 1.0 M HCl with 100 mL of 1.0 NaOH, both at 25 °C. The temperature of the combined solutions and your calorimeter rises by 5.98 °C. The heat capacity of the calorimeter is 100 J/°C. How much energy is released per mol of $H_2O(l)$ formed?

2. An unknown piece of metal weighing 75 g is heated to 95.0 °C. It is dropped into 200 g of water at 25.0 °C. When equilibrium is reached, the temperature of both the water and the piece of metal is 32.0 °C. Determine the specific heat of the metal given that the energy lost by the metal must equal the energy absorbed by the water: $-q_{metal} = q_{water}$. Assume that the heat capacity of the container, a Styrofoam cup, is negligible

Why?

The capability of a chemical system or machine to store and release energy or do work is related to changes in its internal energy. The thermal energy stored or released by a system under constant pressure is called *enthalpy*. The concept of enthalpy will help you better understand the properties of chemical reactions and identify optimum systems for storing and releasing energy and producing work.

Learning Objectives

- Identify the relationships between internal energy, enthalpy, heat, and work
- Learn how to determine changes in internal energy and enthalpy through experimentation

Success Criteria

- Identify processes as endothermic or exothermic
- Determine the sign and magnitude of changes in internal energy and enthalpy

Prerequisite

06-1   *Thermochemistry and Calorimetry*

Information

When physical and chemical changes occur in a system composed of atoms and molecules, energy is usually absorbed or released, and work is often done on or by the system. Energy is required to pull atoms apart and break chemical bonds. When bonds are formed, a more stable and lower energy situation is produced, so energy is released. When a chemical reaction involves bonds breaking **and** being formed, energy is either released or absorbed depending on the balance between the two. The energy change associated with a chemical reaction is usually mostly in the form of thermal energy. In addition to thermal energy or heat, the reaction may produce or absorb energy in the form of work. Work is most easily seen in a chemical reaction that involves gases as reactants or products. As gases are produced say, in the combustion of gasoline in your car's engine, they result in a change in volume, or ΔV. If that expanding gas pushes a piston at constant pressure, the work involved, w, is given up by the reaction system and $w = -P\Delta V$. The energy transferred in a reaction that takes place at a constant pressure is referred to by chemists as the *enthalpy of reaction*. The important relationships among heat, work, enthalpy, and internal energy of a system are represented in **Model 1**.

MODEL   WHAT AFFECTS INTERNAL ENERGY AND ENTHALPY?

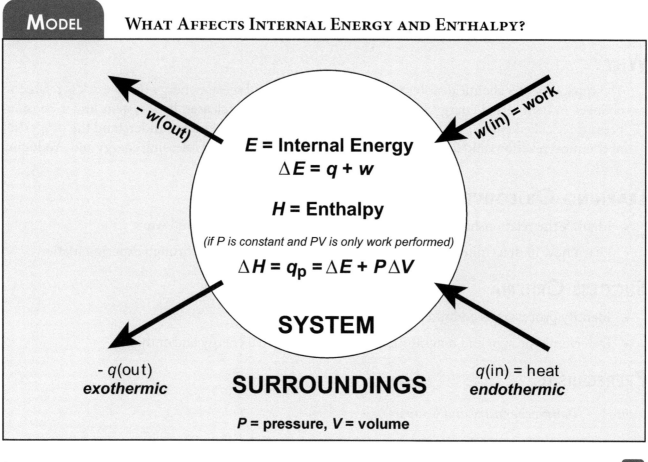

KEY QUESTIONS

1. According to the **Model**, if energy is introduced into a system from the surroundings, how does the internal energy of the system change? What is the sign of this change?

2. According to the **Model**, if work is done on a system (e.g., by compressing it), how does the internal energy of the system change? What is the sign of this change?

3. If a machine such as a steam engine produces work, how does the internal energy of the machine change? What is the sign of this change?

4. Energy flows into or out of a system. What two terms describe the direction of energy flow? What are the definitions of these terms?

5. What is the special name given to the energy (q_p) absorbed or released by a system when the pressure P is constant and the only work done is associated with a volume change?

6. When a gas is compressed at a constant pressure, work is done on the system and its internal energy increases. The work done is given by P times ΔV, where $\Delta V = V_{final} - V_{initial}$. Note that since the gas is compressed, V_{final} is smaller than $V_{initial}$. In calculating the change in internal energy, which of the following expressions should be used? Explain.

$$\text{a) } w = P\,\Delta V \qquad \text{b) } w = -P\,\Delta V$$

7. Since most chemical reactions are conducted in containers open to the atmosphere, why is the energy transfer associated with a chemical reaction generally expressed as a change in enthalpy?

EXERCISES

1. If you do 1 J of work by pulling on a rubber band, by what amount does the internal energy of the rubber band change if there is negligible change in temperature? Write your answer with a positive or negative sign as appropriate.

2. If your hot coffee loses 30 kJ of energy in cooling, what is the change in enthalpy of the coffee? Provide both the sign and the magnitude of ΔH.

3. Under what condition will the changes in enthalpy and internal energy be identical?

4. Identify each of the following as endothermic or exothermic. Explain.

    a. steam condensing

    b. ice melting

    c. gasoline burning

5. Identify whether the sign of ΔH is positive or negative for each of the following. Explain. Change is determined by subtracting the initial value of a quantity from its final value, so:

$$\Delta X = X_{final} - X_{initial}$$

    a. steam condensing

    b. ice melting

    c. gasoline burning

PROBLEMS

1. Your cup of hot coffee loses 50 kJ of energy in cooling, and the volume shrinks because of thermal contraction. For questions i) and ii), select an answer from the following list of possibilities, a) through f), and explain your reasoning.

    a. **exactly** 50 kJ   c. **less** than 50 kJ   e. **more negative** than –50 kJ

    b. **more** than 50 kJ   d. **exactly** –50 kJ   f. **less negative** than –50 kJ

    i. How large is the change in enthalpy of the coffee?

    ii. How large is the change in internal energy of the coffee?

    iii. Justify your answers to i) and ii) by explaining why the change in internal energy must be larger/smaller than the change in enthalpy.

2. Burning butane (C_4H_{10}) produces gaseous carbon dioxide and water. The enthalpy of combustion of butane is – 2650 kJ/mole butane. Determine how much water you can heat from 18 °C to boiling with 1 lb of butane.

3. You are at your cabin in the woods. Will you be able to take a hot bath tonight if only 0.15 lb of propane remains in the tank? Explain. (You will find it necessary to make assumptions or approximations in this problem.) The heat of combustion of propane is –2220 kJ mol^{-1}. The formula for propane is C_3H_8.

WHY?

The conditions that describe a system (such as temperature, pressure, volume, amount of material, and composition) are called its *state*. A quantity is called a *state function* when its value depends only on the state of a material, and not on the path or route used to change it from one state to another. Since enthalpy is a state function, the change in enthalpy for a chemical reaction is the same whether the reaction takes place in one step or in a series of steps. This statement is known as Hess's Law. You can use Hess's Law to find the enthalpy change (ΔH) of a reaction using the ΔH of other, related reactions for which data is already available. Hess's Law extends your ability to evaluate many new chemical reactions for their potential to store and release energy and do work without having to make direct measurements of those reactions.

LEARNING OBJECTIVES

- Understand the concept of a state function
- Understand the definition of standard enthalpy of formation, ΔH_f°
- Master the use of Hess's Law in calculations of reaction enthalpies

SUCCESS CRITERION

- Effectively use Hess's Law in calculations of reaction enthalpies

PREREQUISITE

06-2 *Internal Energy and Enthalpy*

INFORMATION

Hess's Law means that the enthalpy change for any chemical reaction is equal to the sum of the enthalpy changes for any set of other reactions that lead to the same overall change in state. This idea is general and can be applied to any quantity that is a state function. A *state function* is defined as a property that depends only on the state of the system, which means that the property is independent of the history of the system or the path by which the system reached that state.

The *standard enthalpy of formation*, ΔH_f°, for a compound is defined as the change in enthalpy associated with the reaction that creates one mole of that compound (from the requisite elements, in their most stable states and under standard conditions.) These values are found in tables of thermodynamic values, and will specify the states of products when appropriate. For example, ΔH_f° for nitric oxide, NO, implies the reaction:

$$\tfrac{1}{2}\,N_2(g) + \tfrac{1}{2}\,O_2(g) \longrightarrow NO(g)$$

It is difficult to determine through direct experiments the enthalpy of formation of nitric oxide, NO, because the usual product of nitrogen reacting with oxygen is nitrogen dioxide, NO_2, not NO. As illustrated in the following Model, Hess's Law makes it possible to combine the reaction leading to the formation of NO_2 (Reaction 1 in **Model 1**) with the reaction converting NO to NO_2 (Reaction 2 in **Model 1**) in order to obtain the enthalpy of formation of NO.

Since absolute values of enthalpies, H, cannot be measured for compounds, it is necessary to measure changes in enthalpy, $\Delta H = q_p$. In the definition of ΔH_f° for compounds, it is implied that the ΔH_f° for the elements in their standard states is defined as zero. If you were to look up values of ΔH_f° for elements, you would find this to be true.

Reaction 1:	$N_2(g) + 2O_2(g) \rightarrow 2NO_2(g)$	$\Delta H_1 = 68$ kJ
Reaction 2:	$NO(g) + \frac{1}{2} O_2(g) \rightarrow NO_2(g)$	$\Delta H_2 = -56$ kJ
Reaction 3:	$\frac{1}{2} N_2(g) + \frac{1}{2} O_2(g) \rightarrow NO(g)$	$\Delta H_3 = ?$ kJ

If you reverse Reaction 2, you obtain Reaction 4. If you multiply Reaction 1 by ½, you obtain Reaction 5. If you combine Reaction 4 with Reaction 5, you obtain Reaction 3 and a value for ΔH_3.

When you reverse a reaction, you change the sign of ΔH. When you multiply a reaction by a constant, you must multiply ΔH by that same constant. And when you combine or add reactions, you must add the ΔH values together.

Reaction 4:	$NO_2(g) \rightarrow NO(g) + \frac{1}{2} O_2$	$\Delta H_4 = 56$ kJ
Reaction 5:	$\frac{1}{2} N_2(g) + O_2(g) \rightarrow NO_2(g)$	$\Delta H_5 = 34$ kJ
Reactions 4 + 5:	$NO_2(g) + \frac{1}{2} N_2(g) + O_2(g) \rightarrow NO(g) + \frac{1}{2} O_2 + NO_2(g)$	

$$\frac{1}{2} N_2(g) + \frac{1}{2} O_2(g) \rightarrow NO(g)$$

$$\Delta H_3 = \Delta H_4 + \Delta H_5 = -\Delta H_2 + \frac{1}{2} \Delta H_1 = 90 \text{ kJ}$$

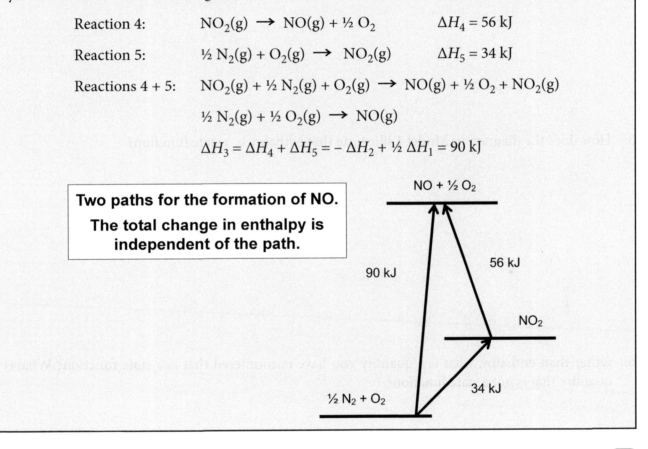

Two paths for the formation of NO.

The total change in enthalpy is independent of the path.

KEY QUESTIONS

1. What is the enthalpy change for the reaction $NO + 1/2\,O_2 \rightarrow NO_2$?

2. What is the enthalpy change for the reaction $NO_2 \rightarrow NO + \frac{1}{2}O_2$?

3. When a reaction is reversed, why does its sign change while its magnitude remains constant?

4. Why is the enthalpy change for Reaction 1 in **Model 1** twice as large as the enthalpy change for Reaction 5?

5. How does the diagram in **Model 1** illustrate that enthalpy is a state function?

6. Other than enthalpy, what is a quantity you have encountered that is a state function? What is a quantity that is not a state function?

EXERCISE

1. Summarize how Reaction 3 and the associated enthalpy change, ΔH_3, are obtained from Reactions 1 and 2 in the Model. Explain why each step is necessary.

MODEL 2	USING HESS'S LAW TO FIND ΔH°_{rxn} FROM ΔH°_f

Based on Hess's Law, the following equation can be used to calculate the standard change in enthalpy for a reaction, ΔH°_{rxn}, from tabulated standard enthalpies of formation, ΔH°_f.

$$\Delta H^\circ_{rxn} = \Sigma m \Delta H^\circ_{f,\,products} - \Sigma n \Delta H^\circ_{f,\,reactants}$$

Where Σ (Greek letter sigma) indicates the sum of all products and reactants and m and n represent the balancing coefficients of each.

For the decomposition of hydrogen peroxide into oxygen and water, the reaction equation is:

$$2H_2O_2(l) \rightarrow 2H_2O(l) + O_2(g)$$

$$\Delta H^\circ_{rxn} = [(2)(\Delta H^\circ_{f,\,H_2O}) + (1)(\Delta H^\circ_{f,\,O_2})] - [(2)(\Delta H^\circ_{f,\,H_2O_2})]$$

The values of the standard enthalpies of formation of these substances can be found in tables and are shown at right:

$\Delta H^\circ_{f,\,H_2O} = -286$ kJ/mol

$\Delta H^\circ_{f,\,O_2} = 0$ kJ/mol, by definition

$\Delta H^\circ_{f,\,H_2O_2} = -188$ kJ/mol

Substituting in these values gives:

$$\Delta H^\circ_{rxn} = [(2)(-286 \text{ kJ/mol}) + (1)(0 \text{ kJ/mol})] - [(2)(-188 \text{ kJ/mol})]$$

$$\Delta H^\circ_{rxn} = \mathbf{-196 \text{ kJ}}$$

KEY QUESTIONS

7. When are ΔH°_f values equal to zero?

8. Why is it necessary to use balancing coefficients in the calculation of enthalpy of reaction?

9. Why is it necessary to subtract the sum of the enthalpies of formation of the reactants from the sum of the enthalpies of formation of the products?

PROBLEMS

1. Calculate the standard enthalpy of formation (ΔH_f°) of ethanol, $C_2H_5OH(l)$, from the heat of combustion of ethanol, which is –1368 kJ/mole, by using standard enthalpies of formation for $CO_2(g)$ $\Delta H_f^\circ = -393.5$ kJ/mol and $H_2O(l)$ $\Delta H_f^\circ = -285.8$ kJ/mol.

2. Draw a diagram, similar to the one in **Model 1**, to illustrate that the enthalpy of formation of gaseous water is different from that of liquid water (both are formed from gaseous dihydrogen and dioxygen.) Indicate on the diagram what that difference is. Include the reaction equations and values for the enthalpy changes at 25 °C in your diagram. You can find enthalpy values in your textbook.

3. Use your diagram from Problem 2 to determine how much energy it takes to convert one mole of liquid water to its gaseous phase at 25 °C.

ACTIVITY 06-4 Chemical Structures and Enthalpy

WHY?

Each reactant and product in a chemical reaction has a unique structure. Enthalpy changes in chemical reactions are therefore related to changes in these chemical structures as the reaction proceeds. This is the reason that enthalpies of formation depend on the identity, structure, and state of each substance. Understanding structure-enthalpy relationships allows estimations of enthalpy changes in new reactions for which there are no enthalpy of formation values. Structure-enthalpy relationships also provide insight into the relative stabilities and reactivities of substances. The structure-enthalpy relationships in two types of compounds you are familiar with, molecular and ionic, will be introduced in this activity. More details about predicting specific molecular structures and enthalpy changes associated with ionic reactions will be covered later in the course.

LEARNING OBJECTIVES

- Understand the concept of *bond energy*
- Understand the series of steps required to form a binary ionic compound from composite elements
- Apply Hess's Law in calculations of reaction enthalpies from structures

SUCCESS CRITERIA

- Effectively calculate reaction enthalpies from molecular structures of reactants and products based on bond energies
- Effectively relate enthalpies of formation of binary ionic compounds with enthalpy terms associated with the formation of gas-phase monatomic ions and *lattice energies*

PREREQUISITES

06-2 *Internal Energy and Enthalpy*

06-3 *Hess's Law: Enthalpy is a State Function*

INFORMATION

When covalent bonds are formed between atoms of non-metals, pairs of electrons are shared between these atoms. As you have seen in the activity on Line Drawings, these bonds can be composed of one pair of electrons, two pairs of electrons, or three pairs of electrons, and each pair is represented with a line between the atoms positions. The number of bonds formed by a pair is called the *bond order*; one pair of atoms can form up to three bonds. Atoms in a covalent bond are more stable than atoms that are unbonded. The enthalpy involved in breaking a covalent bond is always positive; that is, always endothermic. To put it another way, *it always takes energy input to break stable covalent bonds*. These enthalpies for a given type of bond, like a carbon-hydrogen single bond, vary according to the number and types of atoms bonded to the carbon atom, but it is possible to take an average of these individual enthalpies to provide a single value which is called the *bond energy*. Bond energies (BE) can therefore be used to estimate the total enthalpy involved in breaking all the covalent bonds in a set of reactants in a reaction, provided that their structures are known, including the bond order of each bond.

Due to Hess's Law and the fact that enthalpy is a state function, it is not necessary to know the actual reaction path from products to reactants. Therefore, the unlikely reaction that involves breaking all bonds in the reactants, only to rearrange them into new products with new bonds, does not actually have to take place. The enthalpy difference between reactants and products will be the same regardless of the actual process. Since enthalpies of phase changes are not taken into account, bond energy estimates of ΔH°_{rxn} values are most suitable for reactions in the gas phase only.

When using bond energies to estimate the enthalpy change of reaction, the following procedure is often used. All of the bonds in the reactants are broken to form individual atoms. The total enthalpy for this step is the positive sum of all the BE values in all the reactants, taking into account the balancing coefficients for each reactant. On the products side of the reaction equation, the formation of new bonds results in the exothermic release of enthalpy. This process is the reverse of the breaking of bonds, the negative of bond energy for each product can be used to determine the energy released. Overall, the estimated enthalpy change is stated as:

$$\Delta H^{\circ}_{rxn} = \textbf{\textit{sum of all reactant BE values – sum of all product BE values.}}$$

This equation is often confusing for students, because it is similar to—but not quite the same as—the equation used to calculate enthalpies of reaction based on enthalpies of formation. Both give the same end result, but the meaning of bond energy is entirely different from that of enthalpy of formation.

$$\Delta H^{\circ}_{rxn} = \Sigma m \Delta H^{\circ}_{f,products} - \Sigma n \Delta H^{\circ}_{f,reactants}$$

Figure 1 illustrates the difference in the definitions of bond energy and enthalpy of formation.

Figure 1 Difference in the definitions of bond energy and enthalpy of formation

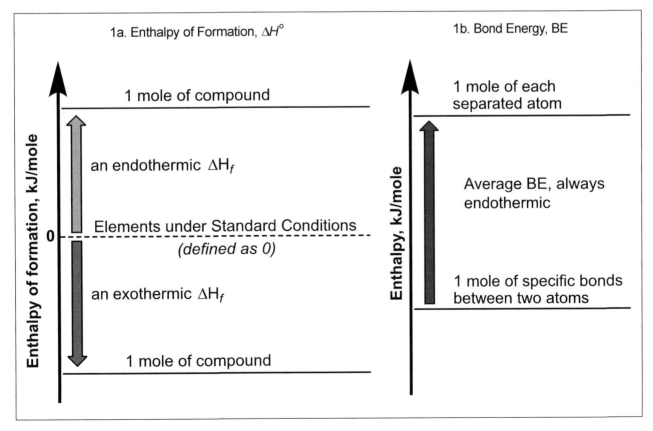

MODEL 1	$\Delta H°_{rxn}$ CALCULATIONS BASED ON BE VALUES VS. $\Delta H°_f$ VALUES

A. Hydrogen peroxide, H_2O_2, is thermodynamically unstable, but requires a catalyst to decompose at a rapid rate according to the reaction:

$$2\ H_2O_2(g)\ \rightarrow\ 2\ H_2O(g) + O_2(g)$$

$$\Delta H°_{rxn} = \boxed{\Sigma m\Delta H°_{f,\ products}} - \boxed{\Sigma n\Delta H°_{f,\ reactants}}$$

$$\Delta H°_{rxn} = \boxed{(2)(\Delta H°_{f,\ H_2O}) + (1)(\Delta H°_{f,\ O_2})} - \boxed{(2)(\Delta H°_{f,\ H_2O_2})}$$

Compound (phase)	$\Delta H°_f$, kJ/mol
$H_2O_2(g)$	–136
$H_2O(g)$	–242
$O_2(g)$	0

$$\Delta H°_{rxn} = \boxed{\begin{array}{l}(2\ mol\ H_2O(g))\ (–242\ kJ/mol) \\ + (1\ mol\ O_2(g))\ (0)\end{array}} - \boxed{(2\ mol\ H_2O_2(g))(–136\ kJ/mol)}$$

$$\Delta H°_{rxn} = \ \textbf{–212 kJ}$$

B. Structural formulas: H–O–O–H, H–O–H, O=O

Bond energy values:

Bond	Bond Energy, kJ/mol
O–H	467
O–O	146
O=O	495

$$H\text{–}O\text{–}O\text{–}H(g)\ +\ H\text{–}O\text{–}O\text{–}H(g)\ \rightarrow\ H\text{–}O\text{–}H(g)\ +\ H\text{–}O\text{–}H(g)\ +\ O{=}O(g)$$

$$\Delta H°_{rxn} = \boxed{\text{sum of all reactant BE values}} - \boxed{\text{sum of all product BE values}}$$

$$\Delta H°_{rxn} = \boxed{\begin{array}{l}(4\ O\text{–}H\ bonds)(BE\ of\ O\text{–}H) \\ + (2\ O\text{–}O\ bonds)(BE\ of\ O\text{–}O)\end{array}} - \boxed{\begin{array}{l}(4\ O\text{–}H\ bonds)(BE\ of\ O\text{–}H) \\ + (1\ O{=}O\ bond)(BE\ of\ O{=}O)\end{array}}$$

$$\Delta H°_{rxn} = \boxed{(2\ O\text{–}O\ bonds)(BE\ of\ O\text{–}O)} - \boxed{(1\ O{=}O\ bond)(BE\ of\ O{=}O)}$$

$$\Delta H°_{rxn} = \boxed{(2\ O\text{–}O\ bonds)(146\ kJ/mol)} - \boxed{(1\ O{=}O\ bond)(469\ kJ/mol)}$$

$$\Delta H°_{rxn} = \ \textbf{–203 kJ}$$

KEY QUESTIONS

1. What chemical reactions are implied by the terms ΔH_f° and **BE** for each of the terms listed in **Model 1**?

ΔH_f°

    $H_2O_2(g)$

    $H_2O(g)$

    $O_2(g)$

BE

    O–H

    O–O

    O=O

2. Write a series of chemical reaction equations corresponding to the implied sequence of chemical steps, or *mechanism*, of decomposition of hydrogen peroxide according to the BE definition.

3. Is this series of steps necessarily the actual mechanism for the reaction? Why or why not?

4. Students often confuse the two equations for calculating ΔH°_{rxn} values from ΔH°_{f} values and BE values. How are these two calculations similar to each other? How are they different?

EXERCISE

1. Using the BE values listed, estimate the heat of combustion of ethanol, $CH_3CH_2OH(g)$.

	Bond energy values			
Bond	**Bond Energy, kJ/mol**		**Bond**	**Bond Energy, kJ/mol**
O–H	467		C–H	413
O–O	146		C–O	358
O=O	495		C=O*	745
C–C	347			* In O=C=O

INFORMATION

When forming binary ionic compounds from metals and non-metals, the solid metal must first be converted into monoatomic metal cations, and the non-metal—which is usually in a polyatomic molecular structure—must first be converted into monoatomic anions. These ions come together through ionic bonding to form alternating three-dimensional structures, with the oppositely charged ions arranged in alternating positions in three dimensions. This rather complicated rearrangement in structure and the transfer of electrons from the metal to the non-metal can be broken down into a series of simple steps with individual enthalpy values that can again be combined using Hess's Law. This enables us to understand each step of the process, along with its relationship to the overall $\Delta H°_{rxn}$ value. This thermochemical cycle is illustrated in a Born-Haber cycle, discussed in **Model 2**.

There are additional enthalpy values in this process that are given special names; you will encounter in more detail when you study trends of properties on the Periodic Table. These new enthalpy terms include the *ionization energy*, symbolized by IE_n, where n designates how many electrons are lost during ionization. IE_1 represents the Ionization Energy of the first electron lost. IE_2 is the Ionization Energy of the second electron lost, etc. These are given in kJ/mole. The enthalpy of the formation of monoatomic gas phase anions from gas phase atoms is represented as *electron affinity*, E_A, also given in kJ/mole. These values are generally determined experimentally. However, the enthalpy of the formation of an ionic solid from gas phase ions cannot feasibly be experimentally determined. The energy of that reaction, called the *lattice energy*, is another enthalpy value that must be calculated using Hess's Law through the Born-Haber cycle. A variety of symbols is used to represent it, but in this presentation, $\Delta H°_{latt}$ will be used.

MODEL 2	ENTHALPY VALUES IN THE FORMATION OF BINARY IONIC COMPOUNDS: THE BORN-HABER CYCLE FOR LiF

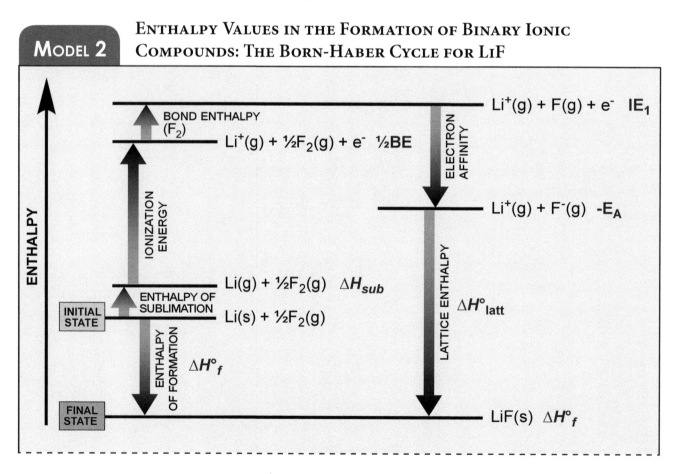

Chapter 6: Energy and Chemical Reactions: Heat and Work

Reactions corresponding to defined enthalpy terms:	$Li(s) + \frac{1}{2}F_2(g) \rightarrow LiF(s)$	$\Delta H_f^\circ = -594.1$ kJ/mole
	$Li(s) \rightarrow Li(g)$	$\Delta H_{sub} = 155.2$ kJ/mole
	$F_2(g) \rightarrow 2\,F(g)$	BE = 150.6 kJ/mole
	$Li(g) \rightarrow Li^+(g) + e^-$	$IE_1 =$ kJ/mole
	$F(g) + e^- \rightarrow F^-(g)$	$E_A = -328$ kJ/mole

Net process in each reaction step of Born-Haber cycle:	$Li(s) + \frac{1}{2}F_2(g) \rightarrow LiF(s)$	$\Delta H_f^\circ = -594.1$ kJ/mole
	$Li(s) + \frac{1}{2}F_2(g) \rightarrow Li(g) + \frac{1}{2}F_2(g)$	$\Delta H_{sub} = 155.2$ kJ/mole
	$Li(g) + \frac{1}{2}F_2(g) \rightarrow Li(g) + F(g)$	1/2 BE = 75.3 kJ
	$Li(g) + F(g) \rightarrow Li^+(g) + F(g) + e^-$	$IE_1 = 520$ kJ
	$Li^+(g) + F(g) + e^- \rightarrow Li^+(g) + F^-(g)$	$E_A = -328$ kJ

Seeking overall reaction for lattice energy:

| $Li^+(g) + F^-(g) \rightarrow LiF(s)$ | $\Delta H_{latt} = ?$ |

Summation of enthalpy reactions and values results in circled reactants and products:	$Li(s) + \frac{1}{2}F_2(g) \rightarrow \boxed{LiF(s)}$	$\Delta H_f^\circ = -594.1$ kJ/mole
	$Li(g) + \frac{1}{2}F_2(g) \rightarrow Li(s) + \frac{1}{2}F_2(g)$	$-\Delta H_{sub} = 155.2$ kJ/mole
	$Li(g) + F(g) \rightarrow Li(g) + \frac{1}{2}F_2(g)$	$-1/2$ BE = 75.3 kJ
	$Li^+(g) + F(g) + e^- \rightarrow Li(g) + F(g)$	$-IE_1 = 520$ kJ
	$\boxed{Li^+(g) + F^-(g)} \rightarrow Li^+(g) + F(g) + e^-$	$-E_A = 328$ kJ

The overall reaction corresponding to unknown Lattice Energy, ΔH_{latt}

| $Li^+(g) + F^-(g) \rightarrow LiF(s)$ | $\Delta H_{latt} = ?$ |

$$\Delta H_{latt} = \Delta H_f^\circ - \Delta H_{sub} - \frac{1}{2}\,BE - IE_1 - E_A$$

$$\Delta H_{latt} = (-594.1) - (155.2) - (75.3) - (520) - (-328)$$

$$\Delta H_{latt} = -1017 \text{ kJ/mol LiF}$$

KEY QUESTIONS

5. Why is it necessary to use so many new enthalpy terms in the Born-Haber cycle for the calculation of ΔH_{latt}?

6. What changes would be required to the Born-Haber cycle in **Model 2** in order to create a Born-Haber cycle for MgF_2?

7. How is lattice energy similar to bond energy? How is it different?

07-1 Electromagnetic Radiation

WHY?

Electromagnetic radiation, which is also called light, is an amazing phenomenon. It carries energy and has characteristics of both particles and waves. We can see only a small region of the electromagnetic spectrum, which we call *visible light*. The absorption and emission of electromagnetic radiation by atoms and molecules serve as powerful tools used to explore molecular structure and chemical reactions. They form the basis of medicine's magnetic resonance imaging and are intrinsic to many analytical techniques used to monitor manufacturing processes and the environment. Radio and television, cell phones, microwave ovens, and compact discs all utilize electromagnetic radiation.

LEARNING OBJECTIVE

- Characterize electromagnetic radiation

SUCCESS CRITERIA

- Interrelate the wavelength, frequency, and energy of electromagnetic radiation
- Identify the different regions of the electromagnetic spectrum

INFORMATION

During the nineteenth century, research in the areas of optics, electricity, and magnetism provided convincing evidence that electromagnetic radiation consists of two oscillating waves. One wave corresponds to an electric field, and the other wave corresponds to a magnetic field. In a vacuum, these fields oscillate perpendicular to each other and perpendicular to the direction the wave is moving. A wave is characterized by its amplitude, frequency, and wavelength. The model in this activity shows a diagram of an electromagnetic wave.

The Greek letter nu, v, is used to represent frequency (cycles/s). Be careful to distinguish it from the English vee, **v**. Frequency is measured in hertz (Hz), which is expressed in cycles or oscillations per second.

The Greek letter lambda, λ, is used to represent wavelength.

During the twentieth century, scientists discovered that electromagnetic radiation also had properties normally associated with particles. This discovery led scientists to believe that electromagnetic radiation consists of particles called photons. A photon has a momentum, a specific amount of energy, and a wavelength and frequency associated with it. Thus, the properties of particles (momentum and a specific energy) and the properties of waves (wavelength and frequency) are blended together.

The wavelength and frequency of electromagnetic radiation extend essentially from 0 to infinity. Scientists have split the spectrum into different "regions," depending on these values. Light from each region of the spectrum interacts with atoms and molecules in characteristic ways, giving rise to absorption or emission of light as these interactions take place. In characterizing the energy levels of electrons in atoms, the ultra-violet, visible, and near-infrared regions are most important. Overall, the divisions between regions are determined by the nature of instrumentation (sources, wavelength selectors, and detectors) used in the different regions and the types of interactions that take place between light and matter. The model in this activity also includes a chart of the electromagnetic spectrum.

MODEL 1 PROPERTIES OF ELECTROMAGNETIC RADIATION

Figure 1 shows an *electromagnetic wave* with the magnetic field oscillating parallel to the *z*-axis, the electric field oscillating parallel to the *y*-axis, and the wave moving along the *x*-axis. The *x*, *y*, and *z*-axes are perpendicular to one other.

The *wavelength* is the distance between any two corresponding points, e.g., from one maximum of the electric field to the next.

The *frequency* is the number of wavelengths that pass a point on the *x*-axis each second.

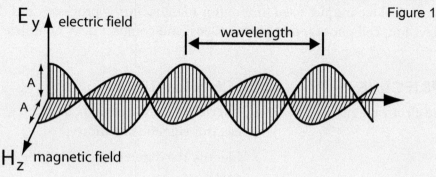

Figure 1

The chart shows the different spectral regions of the electromagnetic spectrum and indicates the approximate frequencies and wavelengths for each. The boundaries between the regions are diffuse.

Figure 2

Frequency (Hz)	Spectral Region	Wavelength (m)
10^{21}	gamma rays	0.3 pm
10^{18}	hard x-rays / soft x-rays	0.3 nm
	vacuum ultraviolet	
10^{15}	ultraviolet / visible	300 nm
	infrared	
10^{12}	far infrared	300 μm
10^{9}	microwave	30 cm
10^{6}	radio wave	300 m

Definitions

v = frequency

λ = wavelength

p = momentum

c = speed of light

$\quad = 2.9979 \times 10^{8}$ m/s

h = Planck's constant

$\quad = 6.6261 \times 10^{-34}$ J s

Photon Properties

energy = $E_{photon} = hv$

momentum = $p = h / \lambda$

frequency = $v = c / \lambda$

1. In the model, what is the equation that shows the relationship between the energy (E) of a photon and its frequency (v)?

2. What is the equation showing the relationship between the frequency (v) and wavelength (λ) of light?

3. What is the equation showing the relationship between the momentum of a photon (p) and its wavelength (λ)?

4. If two waves are traveling at the same speed along the x-axis, will the one with the longer wavelength have the larger or smaller frequency? Explain in terms of the number of wavelengths that pass a given point on the x-axis in 1 second.

5. Is the energy of a photon proportional or inversely proportional to its frequency?

6. Is the momentum of a photon proportional or inversely proportional to its wavelength?

7. Which region of the electromagnetic spectrum has the shortest wavelengths?

8. Which region of the electromagnetic spectrum has photons with the lowest energy?

9. From what regions of the electromagnetic spectrum have you used or encountered photons? Identify the context.

EXERCISES

1. In the model, draw a line connecting two points on the magnetic field wave that are separated by one wavelength.

2. A handheld "red dot" laser pointer that is commonly used to entertain cats (and humans), has a wavelength of 650 nm.

   a. Calculate the frequency of this light.

   b. Calculate the energy of a single photon of this light.

   c. Calculate the momentum of a single photon of this light.

3. Radiation with a wavelength of 100 nm can be used to remove electrons from atoms and molecules. Identify which region of the spectrum corresponds to this radiation.

4. Radiation emitted when excited states of nuclei decay has a frequency of approximately 10^{21} Hz. Identify which region of the spectrum corresponds to this radiation.

5. Identify which uses photons with higher energy: a microwave oven or a radio.

RESEARCH

Radiation is produced and detected in different ways in different regions of the electromagnetic spectrum. Radiation itself has many different applications. If you are familiar with its properties and characteristics, you will be able to assess safety issues, understand the possibilities and limitations of various applications of radiation, and even identify new ones. You can find information on spectroscopy on the internet and in library books on the subject.

Each team should prepare a report to the class on one region of the electromagnetic spectrum. This report should include the following items:

- An object about the size of the radiation's wavelength

- A laboratory source of the radiation

- A method for obtaining tunable monochromatic radiation

- A device that can detect the radiation

- The effect on a molecule when it absorbs the radiation

- Documented effects of the radiation on the human body

- How the radiation is being used in modern research

ACTIVITY 07-2 Atomic Spectroscopy and Energy Levels

Why?

The emission of light by the hydrogen atom (as well as others) played a key role in helping scientists to understand the structure of atoms. The light given off by atoms consists of narrow bands concentrated at specific wavelengths. The graph (or other display) of light intensity as a function of wavelength is called an *emission* or *luminescence spectrum*. The spacing of the energies of the electrons in atoms can be obtained from luminescence spectra.

Learning Objective

- Understand how luminescence spectra can be related to the energy levels of electrons in atoms

Success Criteria

- Calculate the amount of energy absorbed or emitted by a hydrogen atom
- Relate luminescence bands to specific transitions between energy levels

Prerequisites

02-1 *Atoms, Isotopes, and Ions*

07-1 *Electromagnetic Radiation*

Information

A photon is produced when the electrons in an atom lose energy and make the transition from an upper energy level to a lower energy level.

Conservation of energy requires that the energy of the photon (**hv**) must equal the difference in energy between the two levels.

The Model shows the luminescence spectrum of the hydrogen atom in the vacuum ultraviolet region of the electromagnetic spectrum. The intensity of the light emitted (number of photons per second) is plotted on the *y*-axis, and the wavelength in nm is plotted on the *x*-axis.

This series of bands is called the *Lyman series*, after the physicist, Theodore Lyman, who first observed them. They are produced by hydrogen atoms transitioning from excited states (higher energy levels) to the ground state (the lowest energy level).

Each energy level is labeled with an index, **n**, which is also called the *quantum number*. The quantum number **n** has integer values 1, 2, 3, etc. The energy of an energy level is related to its quantum number and is given by the following equation:

$E_n = -2.178 \times 10^{-18} Z^2 / n^2$ Joules, where Z is the atomic number for hydrogen (Z = 1).

The energy values are negative with respect to the ionization threshold (defined as zero). That is, if an electron is promoted to the zero energy level or above, it is ejected from the atom. This simple calculation applies only to atoms with one electron, e.g., the hydrogen atom, H, with Z = 1, or a helium ion, He$^+$, with Z = 2, or a lithium ion, Li^{2+}, with Z = 3.

The energy of a photon involved in a transition or movement of electrons between energy levels is given as the difference of energy levels, final – initial. So, the previous equation becomes

$$E_{photon} = -2.178 \times 10^{-18} Z^2 (1/ n_f^2 - 1/ n_i^2) \text{ Joules}$$

The energy of the photon therefore depends on the *difference* in energy levels given by the principle quantum numbers. If an electron transitions from a higher energy state to a lower energy state, then the energy of the photon is negative, and light is emitted. If the transition is from a lower energy state to a higher energy state, then the energy of the photon is positive and light is absorbed.

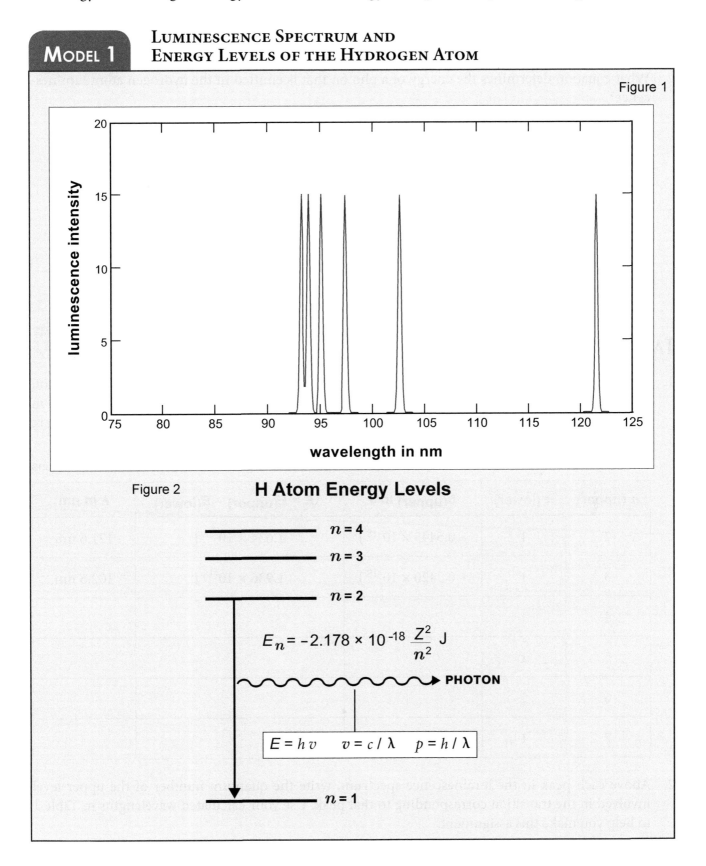

MODEL 1   LUMINESCENCE SPECTRUM AND ENERGY LEVELS OF THE HYDROGEN ATOM

Figure 1

Figure 2

H Atom Energy Levels

$n = 4$

$n = 3$

$n = 2$

$$E_n = -2.178 \times 10^{-18} \frac{Z^2}{n^2} \text{ J}$$

PHOTON

$E = h\nu \qquad \nu = c / \lambda \qquad p = h / \lambda$

$n = 1$

KEY QUESTIONS

1. What is the equation that gives the possible energies (E_n) of an electron in the hydrogen atom?

2. What equation determines the energy of a photon that is emitted in the hydrogen atom luminescence?

3. How can the wavelength of a photon be calculated from its energy?

TASKS

1. In the luminescence spectrum shown in the model, each emitted photon corresponds to a transition between energy levels within the hydrogen atom. Identify these transitions in the model. To aid in this assignment, complete Table 1 below. The first two rows have been completed for you. If you are working in a team of four, each person may do one calculation.

Table 1 Energies of H Atom Levels and Transitions

n (upper)	n (lower)	$E_{(upper)}$ in J	$\Delta E = E_{(upper)} - E_{(lower)}$	λ in nm
2	1	-0.5445×10^{-18} J	1.634×10^{-18} J	121.6 nm
3	1	-0.2420×10^{-18} J	1.936×10^{-18} J	102.6 nm
4	1			
5	1			
6	1			
7	1			

2. Above each peak in the luminescence spectrum, write the quantum number of the upper level involved in the transition corresponding to that peak. Use your calculated wavelengths in Table 1 to help you make this assignment.

3. In the energy-level diagram below (which is based on Figure 2 from the model), draw arrows to represent the transitions corresponding to the spectral bands. The arrow from $n = 2$ to $n = 1$ has been done for you. Label each arrow with the wavelength of the photon produced by that transition. Add additional energy levels to the diagram as needed, and label them with their values for the quantum number n.

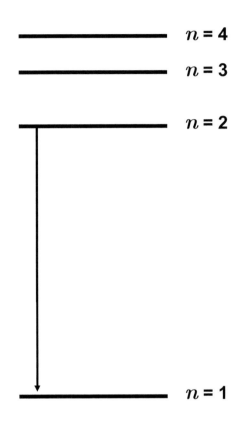

Got It!

1. Compare the transition that occurs when the electron in the hydrogen atom drops from energy level $n = 2$ to $n = 1$ to the transition that occurs from $n = 3$ to $n = 1$.

   a. Which transition causes the larger change in the energy of the hydrogen atom?

   b. Which transition will produce light with a longer wavelength?

2. Why doesn't the hydrogen atom produce light at all wavelengths?

1. a. What is the wavelength of light that is absorbed when an electron in a hydrogen atom goes from the energy level $n = 2$ to $n = 4$?

b. In which region of the spectrum is this wavelength found?

c. Illustrate this transition in an energy level diagram similar to that used in the model.

1. On Earth the ionization energy of atomic hydrogen is 1312 kJ/mol. This is the energy required to remove an electron from an atom scaled up to 1 mole. It will be discussed in detail in a later activity. On another planet the temperature is so high that essentially all the hydrogen atoms have their electron in the $n = 4$ quantum state. What is the ionization energy of atomic hydrogen on that planet in kJ/mole?

2. Calculate the third ionization energy for lithium. The third ionization energy corresponds to the removal of a third electron from the Li^{2+} ion on a per mole basis in kJ/mole.

3. Microwaves are used to heat food in microwave ovens. The microwave radiation is absorbed by moisture in the food. This absorption heats the water, and as it becomes hot, so does the rest of the food. How many photons having a wavelength of 3.00 cm would have to be absorbed by 1.00 g of water to raise its temperature by 1.00 °C?

Quantum Numbers for Electrons in Hydrogen Atoms

WHY?

Electrons behave as three-dimensional stationary waves in atoms and to describe their size, shape, energy and orientation is space it is necessary to use three quantum numbers to describe the orbitals that electrons occupy. The first of these, the principle quantum number, n, was required for hydrogen and hydrogen-like atoms, already discussed, since it is sufficient to explain the spectral behavior of one-electron systems. To consider any atom or ion with more than one electron, we also need to describe the shape of orbitals and their orientation. This activity introduces the quantum number, ℓ, which describes an orbital's three-dimensional shape and the quantum number m_ℓ, which describes an orbital's orientation. This activity will also deal with the hierarchical dependence of the values of these numbers on each other. Comprehension of this content is critical for understanding the structure of the periodic table, the physical and chemical properties of the elements, and chemical bonding theories.

LEARNING OBJECTIVES

- Gain understanding of the existence of atomic orbitals in terms of three quantum numbers
- Recognize the rules governing the hierarchy of quantum numbers

SUCCESS CRITERIA

- Correctly identify particular types of atomic orbitals according to their specific set of three quantum numbers and the relationship between the value of ℓ and the more commonly used letter designation
- Relate characteristics of atomic orbitals, such as their name, shape, orientation in space, nodal pattern, extension in space, and energy

PREREQUISITE

07-2 *Atomic Spectroscopy and Energy Levels*

INFORMATION

The incremental progression in the development of the quantum mechanical model of electrons in atoms, from the initial explanations based on classical mechanics, required several important steps. One of the first was de Broglie's proposal that particles, such as electrons, must have some wave-like properties. Experiments proved this to be true.

This success led Schrödinger to invent a mathematical equation that could be solved to produce *wave functions* describing electrons. This equation is now very famous and is called the *Schrödinger equation*. While the Schrödinger equation is too complicated to be solved exactly for atoms with more than one electron, it can be solved exactly for atoms or ions with a single electron, like the hydrogen atom. The wave functions for the hydrogen atom are used to produce approximate wave functions for all other atoms. These approximate one-electron wave functions are called *atomic orbitals*.

While an atomic orbital is a mathematical function describing a single electron, chemists often think of the orbital as the region of space in which an electron can be found. An orbital can be represented by drawing *a boundary surface*, indicating that the electron has a 90% probability of being within that surface. Boundary surfaces for some atomic orbitals are shown in **Model 1**.

While the energy levels of the hydrogen atom and ions with only one electron are determined by a single index or quantum number, n, Schrödinger discovered that more quantum numbers are involved in determining the energies, shapes, and orientations of the atomic orbitals. In addition to n, the quantum numbers ℓ and m_ℓ are also needed. These quantum numbers are called the *principal quantum number* (n), the *angular momentum* or *azimuthal quantum number* (ℓ), and the *magnetic quantum number* (m_ℓ).

The n quantum number, as in its previous use in describing energy levels in the hydrogen atom emission spectrum, governs the orbital's distance from the nucleus and energy. The value of n dictates the allowed number of values of ℓ associated with it. The maximum value of ℓ is $n - 1$. That is, if $n = 1$, ℓ has only one value, which is 0. If $n = 2$, then the two values of ℓ that are allowed are 0 and 1, since $2 - 1 = 1$.

The value of ℓ is associated with the shape of the orbitals related to it. The primary consideration of orbital shape is the number of *nodal planes* that divide the orbital through the origin. These are also called *angular nodes*. A node is the region within the wave that has an amplitude of zero. You may have been introduced to the concept of nodes as points of zero amplitude in vibrating strings as seen in musical instruments like the guitar. In that case the node is a single point since the string is a one-dimensional object. In orbitals occupying the three dimensional space around the nucleus, the nodal plane(s) bisecting the wave go through the origin defined by the position of the nucleus. In general, ℓ = number of angular nodes describing the orbital.

The values of ℓ are much more commonly referred to as letter names that will dominate any future discussions of it in descriptions of subshells or sublevels on the periodic table, and later in adaptations of atomic orbitals in bonding theories. The correspondence of ℓ values to letter names is given below.

	$\ell = 0$	$\ell = 1$	$\ell = 2$	$\ell = 3$
number of angular nodes	0	1	2	3
shape	spherical	dumbbell	four-leaf clover	complicated with eight lobes
letter designation	s	p	d	f

These letter designations are combined with the principal quantum number to indicate the sublevel or subshell containing one or more orbitals having the same (or very similar) shapes and energies as shown in **Model 1**.

Quantum Numbers n and ℓ Identify Sublevels Having Similar Orbitals

| Symbol | n | | Identifies | "shell" and energy level |

| Name | **Principal** | | Allowed Values | n = 1, 2, 3, … to infinity |

| Characterizes | **size energy** total nodes = $n - 1$ |

n = number of ℓ values allowed in the shell

n^2 = total possible number of orbitals in the shell

| Symbol | ℓ | | Identifies | "subshell" and energy sublevel |

| Name | **Angular Momentum** *or* **Azimuthal** | Allowed Values | ℓ = 0, 1, 2,…, $(n - 1)$ |

| Characterizes | **shape** |

energy in multi-electron atoms angular nodes which are number of planar nodes = ℓ

Table 1

Allowed sublevel designation

		ℓ value			
		0	1	2	3
n value	1	1s			
	2	2s	2p		
	3	3s	3p	3d	
	4	4s	4p	4d	4f
	5	5s	5p	5d	5f
	6	6s	6p		

Figure 1

Characteristic orbital shapes and corresponding sublevel names based on values of n and ℓ (Note that the differences in orbital shapes and the relative sizes of the s orbitals are our only considerations at this point.)

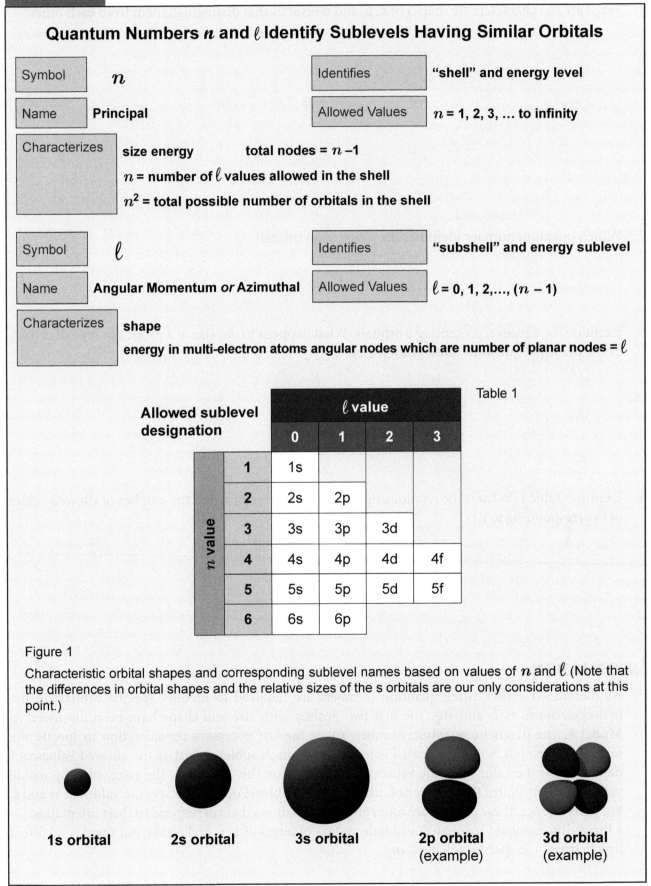

1s orbital   **2s orbital**   **3s orbital**   **2p orbital** (example)   **3d orbital** (example)

KEY QUESTIONS

1. What are the characteristic shapes of **s**, **p**, and **d** orbitals that distinguish them from each other?

2. Which quantum number identifies the shape of an orbital?

3. Examine the figures representing **s** orbitals. What happens to the size of a particular type of orbital, as the principal quantum number increases?

4. Examine Table 1. What is the relationship between the value of n and the number of allowed values of ℓ corresponding to it?

INFORMATION

As mentioned earlier, three quantum numbers are required to identify specific orbitals in the hydrogen atom, n, ℓ, and m_ℓ. The first two dealing with size and shape have been discussed in **Model 1**. The magnetic quantum number, m_ℓ, is the last necessary consideration to specify the number of orbitals, and their spatial orientations in each sublevel. Just as the allowed values of ℓ depended on the value of n, the values of m_ℓ depend on the value of ℓ. The range of the possible values of m_ℓ indicates the number of orbitals in the sublevel defined by specific values of n and ℓ. For convenience, these orbitals are often given designations that correspond to their orientation on a three-dimensional Cartesian coordinate system in terms of x, y, and z axes, but there is no direct link between the specific values of m_ℓ.

Chapter 7: Electron Configurations and the Periodic Table

THE MAGNETIC QUANTUM NUMBER, m_ℓ
IN HYDROGEN ATOMIC ORBITALS

Symbol	m_ℓ		Identifies	orbitals in a "subshell"
Name	**Magnetic**		Allowed Values	$m_\ell = -\ell, (-\ell + 1), \dots 0, \dots (\ell - 1), +\ell$
Characterizes	orientation			

The Number of Orbitals in a Subshell (s, p, or d) and their Possible Orientations in Space Around the Nucleus of the Hydrogen Atom

Figure 2

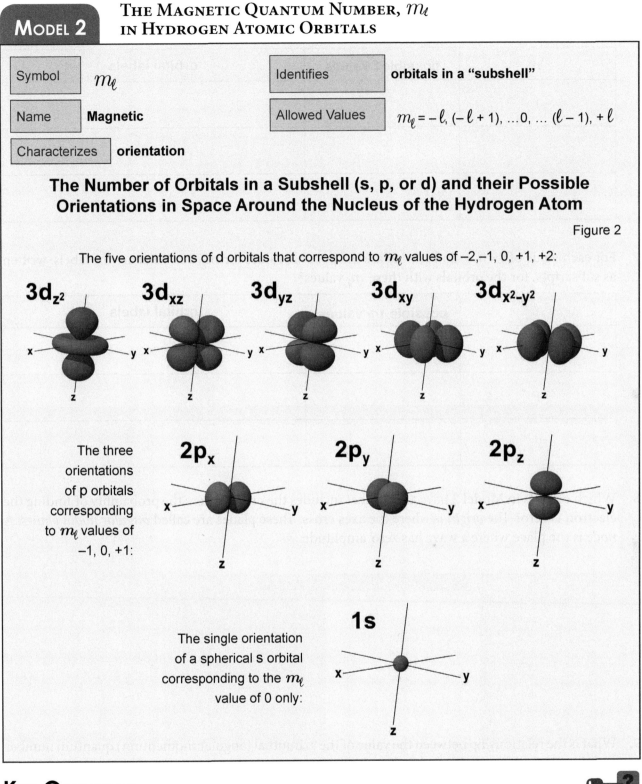

The five orientations of d orbitals that correspond to m_ℓ values of −2, −1, 0, +1, +2:

$3d_{z^2}$ $3d_{xz}$ $3d_{yz}$ $3d_{xy}$ $3d_{x^2-y^2}$

The three orientations of p orbitals corresponding to m_ℓ values of −1, 0, +1:

$2p_x$ $2p_y$ $2p_z$

The single orientation of a spherical s orbital corresponding to the m_ℓ value of 0 only:

1s

KEY QUESTIONS

5. According to the models, how many orbitals correspond to each of the following values of n ?

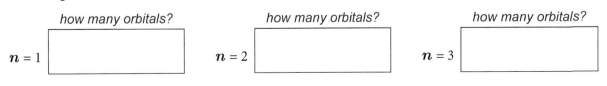

how many orbitals?

$n = 1$ [ ]

how many orbitals?

$n = 2$ [ ]

how many orbitals?

$n = 3$ [ ]

6. For each value of n = 1, 2, and 3, what are the possible values for ℓ, and what are the labels for the orbitals with these ℓ values? (For example s, p, or d.)

n	possible ℓ values	orbital labels
1		
2		
3		

7. For each value of ℓ = 0, 1, and 2, what are the possible values for m_ℓ, and what are the labels, written as subscripts, for the orbitals with these m_ℓ values?

ℓ	possible m_ℓ values	orbital labels
0		
1		
2		

8. Which orbitals in **Model 2** have a plane that includes the origin where the probability of finding the electron is zero? The *origin* is where the axes cross. These planes are called *angular nodal planes*. A *node* is the place where a wave has zero amplitude.

9. What is the relationship between the value of the azimuthal (angular momentum) quantum number and the number of angular nodal planes?

Chapter 7: Electron Configurations and the Periodic Table

INFORMATION

The idea of angular nodal planes is essential to the description of the shapes of orbitals determined by the value of ℓ, however there is an additional concept that provides further insight into the wave functions of hydrogen atomic orbitals. Orbitals having different principal quantum numbers, n, but identical angular momentum quantum numbers, ℓ, are similar in overall shape, but not identical. It has already been stated that as the value of n increases, the size of the orbital increases, but in addition to this difference is the appearance of *radial nodes* on the interior regions of the orbitals. These nodes may be a little harder to imagine, but are *spherical nodes* that exist at specific distances from the nucleus. When a subshell first appears in building up orbitals based on values of n, for example the three orbitals in the 2p subshell, the only node in each is the angular node.

No radial nodes would occur in the orbitals of the a 2p subshell. However, as the value of n increases beyond 2, radial nodes occur in the three orbitals so that the total number of nodes in each orbital continues to be $n - 1$. So, by comparison, a 3p orbital also has one planar angular node since $\ell = 1$. But, since $n - 1 = 2$, a second node, which is a radial node, must be present within each of the 3p orbitals.

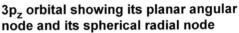

$3p_z$ orbital showing its planar angular node and its spherical radial node

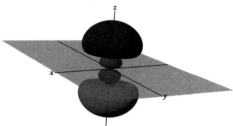

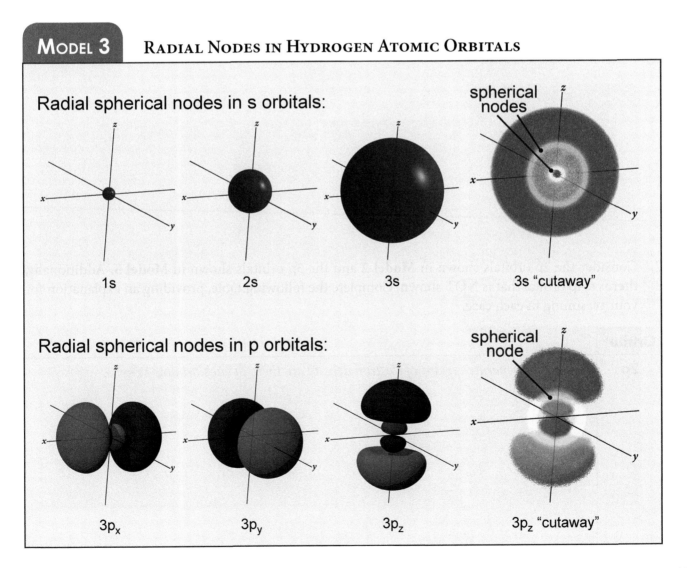

MODEL 3   **RADIAL NODES IN HYDROGEN ATOMIC ORBITALS**

Radial spherical nodes in s orbitals:

1s   2s   3s   spherical nodes   3s "cutaway"

Radial spherical nodes in p orbitals:

$3p_x$ $3p_y$ $3p_z$ spherical node $3p_z$ "cutaway"

KEY QUESTION

10. Consider the 1s, 2s, and 3s orbitals in Model 3. Complete the following table.

Orbital	Number of regions of electron density	Number of radial nodes
1s		
2s		
3s		

EXERCISES

1. The total number of nodes in an orbital is equal to $n - 1$. This number is split between radial nodes and angular nodal planes. Show that your answers to Key Question 10 are consistent with this with fact.

2. Consider the 2p orbitals shown in **Model 2** and the 3p orbitals shown in **Model 3**. Additionally, there's a 4p orbital that is NOT shown. Complete the following table, providing an explanation for your reasoning in each case.

Orbital	
2p	*How many separate regions of electron density are there in each orbital? Why?*

Chapter 7: Electron Configurations and the Periodic Table

2p	*How many radial nodes are there in each orbital? Why?*

3p	*How many separate regions of electron density are there in each orbital? Why?*
	How many radial nodes are there in each orbital? Why?

4p	*How many separate regions of electron density are there in each orbital? Why?*
	How many radial nodes are there in each orbital? Why?

MODEL 4   ENERGY LEVELS OF HYDROGEN ATOMIC ORBITALS

$$0$$

$\ell = 0$ $\ell = 1$ $\ell = 2$

$m_\ell = 0$ $m_\ell = 0$ $1,-1$ $1,-1$ $m_\ell = 0$ $1,-1$ $1,-1$ $2,-2$ $2,-2$

$-E_0/9$ $n = 3$ $\underline{3s}$ $\underline{3p_z}$ $\underline{3p_x}$ $\underline{3p_y}$ $\underline{3d_{z^2}}$ $\underline{3d_{xz}}$ $\underline{3d_{yz}}$ $\underline{3d_{xy}}$ $\underline{3d_{x^2-y^2}}$

$-E_0/4$ $n = 2$ $\underline{2s}$ $\underline{2p_z}$ $\underline{2p_x}$ $\underline{2p_y}$

$E \uparrow$

$-E_0$ $n = 1$ $\underline{1s}$

For atoms or ions with only 1 electron, orbital energies are ordered as follows:
$$1s < 2s = 2p < 3s = 3p = 3d < 4s = 4p = 4d = 4f < 5s = 5p = 5d = 5f < 6s = 6p = 6d < 7s$$

KEY QUESTIONS

11. How are the energy levels shown in **Model 4** consistent with the information in Activity 7-2?

1. In a hydrogen atom with one electron in the excited state of $n = 3$, which orbital is occupied by the electron? Why?

2. From the pattern in **Model 4**, develop a general statement about the relationship between the total number of nodes in orbitals and their relative energies.

ACTIVITY 07-4 Multi-Electron Atoms, the Aufbau Principle, and the Periodic Table

WHY?

To construct the other elements from the simplest element, hydrogen, protons and neutrons are added to the nucleus and electrons are added to the orbitals in order of increasing energy. This is called the building-up principle (*aufbau* in German). If you know an element's orbital energy-level structure and the number of electrons it has, you can determine the electron configuration and thus the properties of the element. This information is summarized in the *Periodic Table of the Elements*, which you can use as a tool to identify materials with similar or contrasting chemical and physical properties.

LEARNING OBJECTIVE

- Master the procedure for determining the electron configuration of atoms and ions

SUCCESS CRITERIA

- Correctly write the electron configuration of atoms and ions
- Relate electron configuration to positioning in the Periodic Table

PREREQUISITES

07-2   *Atomic Spectroscopy and Energy Levels*

07-3   *Quantum Numbers for Electrons in Hydrogen Atoms*

MODEL 1   ENERGY LEVEL DIAGRAM FOR THE HYDROGEN ATOM

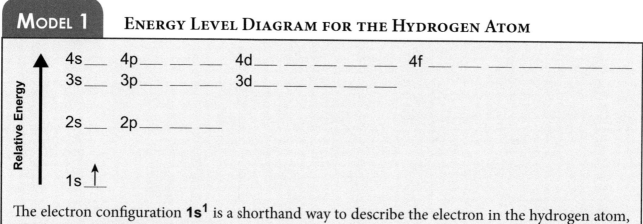

The electron configuration **1s^1** is a shorthand way to describe the electron in the hydrogen atom, which is represented by an arrow.

KEY QUESTIONS

1. In the electron configuration for hydrogen, 1s^1,

   a. What does the first 1 refer to?

With contributions by Vicky Minderhout and colleagues at Seattle University

b. What does the **s** refer to?

c. What does the superscript 1 refer to?

2. Why are there different numbers of lines drawn after the symbols; e.g., **4s** has one line, **4p** has three lines, **4d** has five lines and **4f** has seven lines?

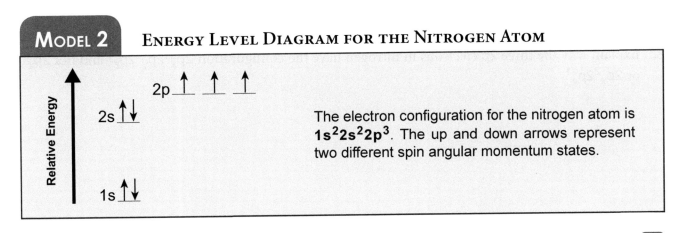

MODEL 2 — ENERGY LEVEL DIAGRAM FOR THE NITROGEN ATOM

The electron configuration for the nitrogen atom is $1s^2 2s^2 2p^3$. The up and down arrows represent two different spin angular momentum states.

KEY QUESTIONS

3. Comparing **Model 1** and **Model 2**, what happens to the energies of the 2s and 2p orbitals when the atom has more than one electron?

4. In the notation $2p^3$, what information is provided by the 2, the p, and the 3?

5. What information about the electron is provided by the direction of the arrow in the diagram?

6. In nitrogen, are two electrons with the same spin angular momentum described by the same atomic orbital? Explain your answer. The *Pauli Exclusion Principle* explains your observation.

7. If two atomic orbitals have the same energy, will two electrons pair up in one of the two or enter different orbitals? *Hund's Rule* explains your observation.

8. Explain why the three 2p electrons in nitrogen have the configuration $2p_x^1 2p_y^1 2p_z^1$ and not $2p_x^3$ or $2p_x^2 2p_y^1$.

INFORMATION

All atoms have the same kinds of orbitals as the hydrogen atom, but the energies of these orbitals change as the number of protons and the number of electrons in an atom changes. The following guidelines are used to identify the orbitals that are occupied by electrons in any atom. These guidelines are derived from quantum mechanics and the solutions to Schrödinger's equation.

The *Aufbau Principle* says that, as protons are added to the nucleus to build up elements, electrons are always added to (available) lower energy orbitals before they fill higher energy orbitals. Hence, *aufbau*: building-up. The Aufbau Principle makes sense because it produces lowest-energy, or most stable, arrangement of the electrons in the atom.

The *Pauli Exclusion Principle* says that two electrons cannot have the same set of quantum numbers simultaneously. This principle determines the number of electrons that can occupy each orbital. In addition to the three quantum numbers introduced in *Activity 07-3: Quantum Numbers for Electrons in Hydrogen Atoms*, there is a fourth quantum number that describes angular momentum. Electron spin angular momentum is a property that is described by Dirac's relativistic formulation of quantum mechanics. In some ways it is similar to a basketball or tennis ball spinning; however there are important differences. Electron spin angular momentum is best regarded as the intrinsic angular momentum carried by an electron and not as the electron actually spinning. Dirac showed that each electron can have two values for the spin angular momentum quantum number, m_s: $+\frac{1}{2}$ or $-\frac{1}{2}$. Therefore, two electrons can be in each orbital and have the same values for n, ℓ, and m_ℓ, as long as they have different values for m_s. The two different spin angular momentum states for electrons are represented in orbital diagrams by up and down arrows, or by the values $+\frac{1}{2}$ and $-\frac{1}{2}$ for the spin quantum number, m_s.

Hund's Rule says that if multiple orbitals with the same energy are available, then the unoccupied orbitals will be filled by electrons with the same spin before electrons with different spins pair up in occupied orbitals. Hund's Rule makes sense because electrons repel each other. If they are in the same orbital, they will be close together, and their energy will be higher than it would be if they were separated in different orbitals.

EXERCISES

1. Complete the energy level diagram below by filling in the electron configuration for sulfur.

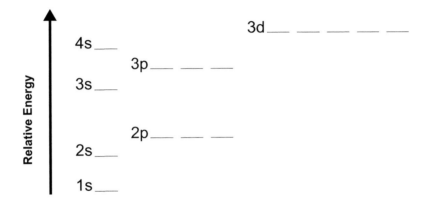

2. In your energy level diagram for sulfur, above, identify features that illustrate

   a. the Aufbau Principle

   b. the Pauli Exclusion Principle

   c. Hund's Rule.

3. Complete the energy level diagram below and write the electron configuration for iron.

3d ___ ___ ___ ___ ___

4s ___

3p ___ ___ ___

3s ___

Relative Energy

2p ___ ___ ___

2s ___

1s ___

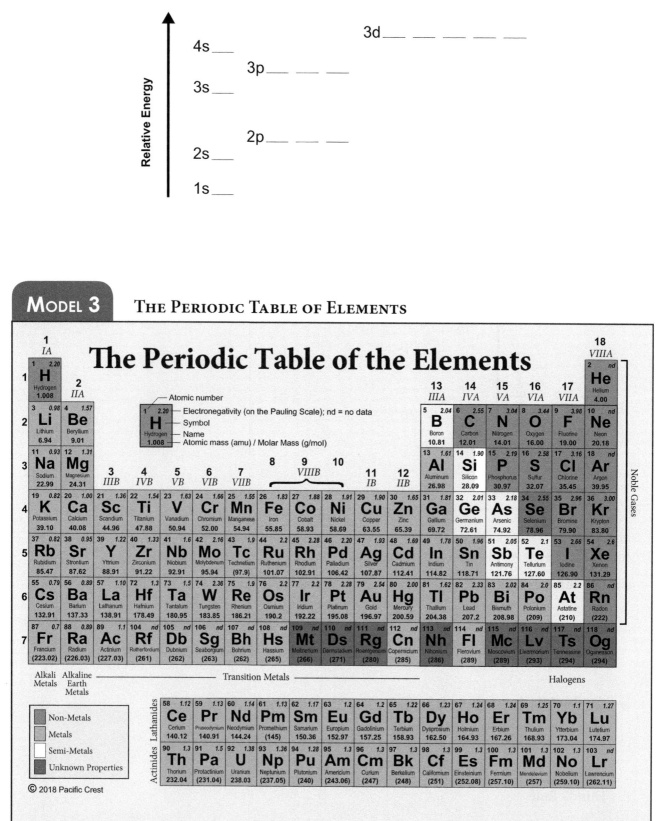

MODEL 3   **THE PERIODIC TABLE OF ELEMENTS**

The rows in the Periodic Table are called **periods**, the columns are called **groups**, and the elements can be grouped together in blocks. All of this structure reflects the electron configurations of atoms. The information in the Electron Configuration Table (following) and the following Key Questions should help you deepen your understanding of the Periodic Table.

Electron Configurations

Table 1

Element	Number of Electrons	Electron Configuration	Element	Number of Electrons	Electron Configuration
He	2	$1s^2$	K	19	$[Ar]4s^1$
Li	3	$1s^2 2s^1$	Sc	21	$[Ar]4s^2 3d^1$
Be	4	$1s^2 2s^2$	Ti	22	$[Ar]4s^2 3d^2$
B	5	$1s^2 2s^2 2p^1$	Zn	30	$[Ar]4s^2 3d^{10}$
Ne	10	$1s^2 2s^2 2p^6$	Ga	31	$[Ar]4s^2 3d^{10} 4p^1$
Na	11	$[Ne]3s^1$	Kr	36	$[Ar]4s^2 3d^{10} 4p^6$
Al	13	$[Ne]3s^2 3p^1$	Rb	37	$[Kr]5s^1$
Ar	18	$[Ne]3s^2 3p^6$			

KEY QUESTIONS

9. Keeping electron configurations in mind, why there are only two elements in Period 1?

10. Again considering electron configurations, why do Periods 2 and 3 each contain eight elements?

11. Why are there 18 elements in Period 4?

12. A block is formed by all the elements in Groups 1 and 2. What do these elements have in common?

13. A block is formed by all the elements in Groups 3 through 12. What do these elements have in common?

14. A block is formed by all the elements in Groups 13 through 18. What do these elements have in common?

15. A block is formed by the 28 elements set off at the bottom of the table. What do these elements have in common?

16. Why are the chemical properties of Na, K, and Rb similar?

17. Considering their electron configurations, why do you think the elements in Group 18 are very stable, inert, and unreactive?

18. What do the elements in each group have in common?

4. Identify three elements not in Groups 1 or 18 that you expect to have similar chemical properties. Explain why.

5. Identify the element with the following electron configuration: $[Xe]6s^24f^{14}5d^{10}$.

6. Write the electron configuration for each of the following atoms or ions:

O   _____

Mg   _____

Mg^{2+} _____

Cl   _____

Cl$^-$ _____

N^{3-} _____

7. What change in the number of electrons is necessary in forming cations?

ACTIVITY 07-5 Periodic Trends in Atomic Properties

Why?

Many properties of atoms have a repeating pattern when plotted with respect to atomic number. These similarities are due to the repeating pattern of electron configurations involving s, p, d, and f orbitals. These experimentally observed periodic trends actually provide evidence for the orbital and shell structure of atoms, as well as the meaningful arrangement of elements in the Periodic Table. These atomic-level chemical and physical properties are important for understanding the properties of the elements at the laboratory level as well. Your ability to recognize the properties of the elements from their positions in the Periodic Table will prove useful in handling chemical compounds safely, developing new materials, finding new applications of known materials, or using chemistry in medical applications.

Learning Objective

- Understand relationships between position in the Periodic Table, electron configurations, atomic radius, ionization energy, and electron affinity

Success Criteria

- Use electron configurations and position in the Periodic Table to account for the relative sizes, ionization energies, and electron affinities of different atoms

- Order elements by atomic radius, ionization energy, and electron affinity based on observed trends

Prerequisites

07-3 *Quantum Numbers for Electrons in Hydrogen Atoms*

07-4 *Multi-electron Atoms, the Aufbau Principle, and the Periodic Table*

Information

Properties such as the size of an atom (atomic radius), the energy required to remove an electron from an atom in the gas phase (ionization energy), and the energy involved an adding an electron to an atom to form a negatively charged atomic anion (electron affinity) can be understood in terms of the electron configuration of the atom and the balance between *electron-nucleus attraction* and *electron-electron repulsion*. Ionization energies, which are always positive (endothermic), are usually given in units of kJ/mol. Electron affinities, which are usually negative (exothermic), are also reported in units of kJ/mol. An electron in an atom is attracted by the positively charged nucleus and repelled by the other electrons. Atomic configuration depends on the relative strengths of these interactions. If the electron-nucleus attraction (e-n) has a large effect, then the atom is small, has a high, positive ionization energy, and has a large, negative electron affinity. If the electron-electron repulsion (e-e) has a large effect, then the atom is large, the ionization energy is small, and the electron affinity is very small, zero, or even positive. Table 1 summarizes what happens in the progression from one atom to the next in the Periodic Table; both the nuclear charge and the number of electrons increase by 1 unit.

Table 1

Interaction	Atomic Size	Ionization Energy	Electron Affinity
e-n attraction > e-e repulsion	Decreases	Increases	Increases
e-n attraction < e-e repulsion	Increases	Decreases	Decreases

MODEL 1 Review of Electron Configurations of Atoms

The electron configuration of an atom specifies the number of electrons in each atomic orbital.

Example: The Electron Configuration of Carbon

$$1s^22s^22p^2$$

This notation for the electron configuration means that 2 electrons are in the 1s orbital, 2 are in the 2s orbital, and 2 are in 2p orbitals. Electron configurations can also be represented by orbital box diagrams, as shown below. In these diagrams, an electron is represented by an upward or a downward pointing arrow. An upward pointing arrow indicates that the electron has positive spin angular momentum in one direction (referred to as *spin up*). A downward pointing arrow indicates that it has negative spin angular momentum in that direction (referred to as *spin down*). Dirac showed that the idea of spin angular momentum is a consequence of Einstein's theory of special relativity and does not mean that the electron is spinning like a top.

The following Key Questions will help you identify the principles or rules that account for electron configurations by helping you examine the orbital box diagrams in Figure 1.

Figure 1

	1s	2s	$2p_x$	$2p_y$	$2p_z$
H	↑				
He	↑↓				
Li	↑↓	↑			
Be	↑↓	↑↓			
B	↑↓	↑↓	↑		

	1s	2s	$2p_x$	$2p_y$	$2p_z$
C	↑↓	↑↓	↑	↑	
N					
O					
F					
Ne					

Key Questions

1. Consider the orbital box diagrams in **Model 1**. Do electrons fill the lower or higher energy orbitals first?

2. What is the maximum number of electrons that can fit in each orbital?

3. If multiple orbitals with the same energy are open, e.g., in the case of 2p orbitals, do two electrons fill one open orbital, or do they enter different orbitals?

4. According to the information in **Model 1**, how many different spin angular momentum states are there for each electron?

5. What are two ways shown in **Model 1** to write or specify the electron configuration of an atom?

6. What are the names of the three guidelines that determine the electron configuration of an atom?

7. In the orbital box diagram for carbon shown in **Model 1**, why aren't both p-electrons in the p_x-orbital?

8. Why isn't the electron configuration of carbon $1s^3 2s^3$?

Chapter 7: Electron Configurations and the Periodic Table

EXERCISES

1. Complete the orbital box diagrams in **Model 1** by applying the Aufbau Principle, the Pauli Exclusion Principle, and Hund's Rule.

MODEL 2 — ELECTRON-ELECTRON AND ELECTRON-NUCLEUS INTERACTIONS IN ATOMS

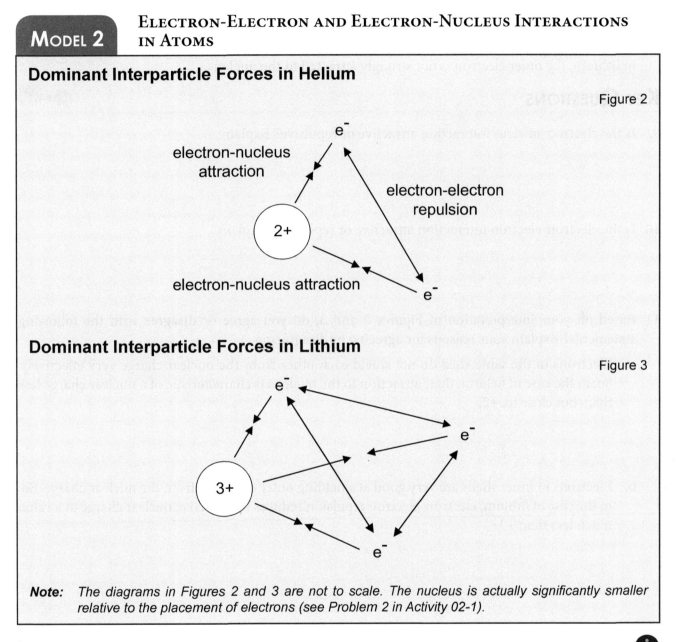

Dominant Interparticle Forces in Helium

Figure 2

Dominant Interparticle Forces in Lithium

Figure 3

Note: The diagrams in Figures 2 and 3 are not to scale. The nucleus is actually significantly smaller relative to the placement of electrons (see Problem 2 in Activity 02-1).

INFORMATION

The diagrams in **Model 2** illustrate how electron-electron repulsion can effectively reduce the attraction of an electron to the nucleus, and how this *shielding effect* depends on the electron configuration.

In helium, both electrons are in the same atomic orbital (1s), and therefore do not shield each other very well from the +2 charge of the nucleus. Consequently, the two electrons are strongly attracted to the nucleus.

Effectiveness of shielding also depends on orbital shape, due to the existence of nodes. For example, a 1s orbital, with no angular or radial nodes, has high electron density close to the nucleus and therefore shields very effectively. A **2p** orbital, with an angular node, has no electron density at the nucleus and therefore shields less effectively. Moving across a period from left to right, the addition of electrons to the p orbital reduces the effectiveness of shielding, giving rise to greater (n-e) attractive forces, increasing ionization energies, and making electron affinities more negative.

In lithium, the outer electron is in a **2s** orbital and is shielded very effectively from the +3 charge of the nucleus by the two electrons in the inner **1s** orbital. As a result of this shielding and because it is farther from the nucleus, the outer electron experiences a smaller nuclear charge. Consequently, in lithium, the outer electron is not strongly attracted to the nucleus.

KEY QUESTIONS

9. Is the electron-nucleus interaction attractive or repulsive? Explain.

10. Is the electron-electron interaction attractive or repulsive? Explain.

11. Based on your interpretation of Figures 2 and 3, do you agree or disagree with the following statements? Explain your reasons for agreeing or disagreeing.

    a. Electrons in the same shell do not shield each other from the nuclear charge very effectively. So, in the case of helium, their attraction to the nucleus is characteristic of a nuclear charge less than, but close to, +2.

    b. Electrons in inner shells are very good at shielding outer electrons from the nuclear charge. So, in the case of lithium, electron-electron repulsion reduces the effective nuclear charge to a value much less than +3.

INFORMATION

Ionization energy is defined as the energy required to ionize an atom and is defined by the following reaction equation. All species are in the gas phase. (Ionization energy is a positive quantity.)

$$X \rightarrow X^+ + e^-$$

Electron affinity is the energy required to add an electron to an atom to form an atomic anion. It is defined by the following reaction equation. Again, all species are in the gas phase. (Electron affinity is usually a negative quantity.)

$$Y + e^- \rightarrow Y^-$$

Chapter 7: Electron Configurations and the Periodic Table

12. Will an increase in electron attraction to the nucleus tend to increase or decrease each of the following? Explain.

   a. The size of an atom

   b. The ionization energy

   c. The electron affinity

13. Will an increase in the repulsion of an electron by other electrons tend to increase or decrease each of the following? Explain.

   a. The size of an atom

   b. The ionization energy

   c. The electron affinity

EXERCISES

2. In going from hydrogen to helium, there is a change in the atomic radius (from 37 to 32 pm) and a change in ionization energy (from 1311 kJ/mole to 2377 kJ/mole). Identify what these changes suggest about the relative magnitudes of the changes in the *electron–nucleus attraction* and the *electron–electron repulsion*.

3. Using the ideas illustrated in **Model 2** and developed by the preceding Key Questions and Exercise 2, explain why you would expect lithium to have a larger atomic radius and a smaller ionization energy than helium.

MODEL 3 — VARIATION OF ATOMIC PROPERTIES WITH ATOMIC NUMBER

The unit for atomic radii is pm; the unit for ionization energies and electron affinities is kJ/mole.

Figure 4

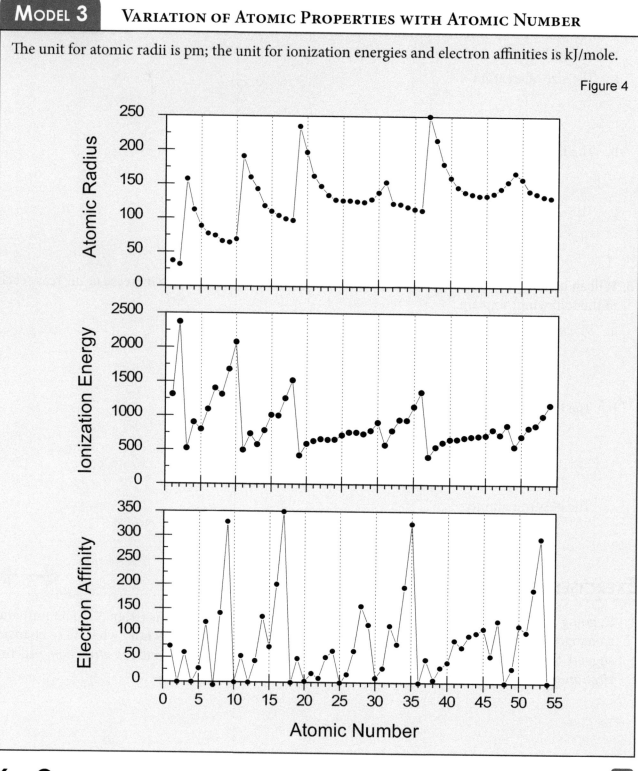

KEY QUESTIONS

14. The Periodic Table is organized with the atomic number of the elements increasing from left to right and top to bottom. Similarly, the *x*-axis for all graphs in **Model 3** is the atomic number of each plotted element. When going from one element to the next across the *x*-axis, what happens to the number of protons and the number of electrons?

15. Do additional protons tend to increase or decrease the electron-nucleus attraction? Explain.

16. Do additional electrons tend to increase or decrease the electron-electron repulsion? Explain.

17. Using the information from the graphs in **Model 3**, describe what happens to the atomic radius and ionization energy as you go across a row in the Periodic Table, e.g., from Li to Ne and Na to Ar?

18. Why do the trends that you identified in Key Question 17 occur? In your explanation, use the increase in nuclear charge and the effectiveness of electron shielding by electrons in the same orbital, as illustrated in **Model 2**.

19. Using the information from the graphs in **Model 3**, describe what happens to the atomic radius and ionization energy when going down a group in the Periodic Table, e.g., from Ne to Xe or Li to Rb.

20. Why do the trends that you identified in Key Question 19 occur? In your explanation, use the increase in nuclear charge, the effect of electron shielding of the nuclear charge by electrons in inner shells, and the size of the outer shell.

21. How can you use the Periodic Table and electron configurations to predict relative atomic radii and ionization energies for two atoms?

22. Summarize trends in atomic radius, ionization energy, and electron affinity in the outline sketches of the Periodic Table shown below. Do this by identifying, **with a single arrow**, the following trends in each sketch:

        a. the trend for decreasing atomic radius

        b. the trend for increasing ionization energy

        c. the trend for increasingly negative electron affinities

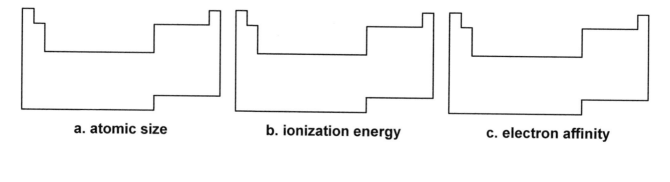

      a. atomic size   **b. ionization energy**   **c. electron affinity**

EXERCISES

4. Identify which of the two atoms in each pair is larger, and explain why.

    a. calcium or potassium

b. nitrogen or oxygen

c. sodium or potassium

5. Identify which of the elements in each pair has the higher ionization energy, explaining why for each.

a. Ba or Cs

b. Br or Kr

c. Si or C

PROBLEMS

1. In the Periodic Table, ionization energies tend to increase across a period or row. For the ionization energies in the second period (Li to Ne), identify where exceptions to this general trend occur. To help you explain these exceptions, use orbital box diagrams like those in **Model 1** and your knowledge of the differences in electron configurations between neighboring atoms.

2. Using the insight you gained from Problem 1, explain:

   a. Why the first ionization energy of sulfur is less than the first ionization energy of phosphorus.

b. Why the first ionization energy of aluminum is less than the first ionization energy of magnesium.

3. Variations in electron affinities can be understood in the same way as variations in ionization energies except that, in the analysis of the effects of electron shielding and nuclear charge, one uses the electron configuration and the orbital box diagram for the anion rather than for the atom. Using this idea, identify the atom in each pair that has the greater electron affinity and explain why.

a. H or He

b. He or Li

c. B or C

d. Cl or Ar

4. Explain why the electron affinity of Be is very small, or zero, and the electron affinity of F is very large (328 kJ/mole).

5. Why do peaks in ionization energy occur for atoms with atomic numbers equal to 10, 18, and 36?

6. Elements with atomic numbers equal to 9, 17, and 35 have large electron affinities, but those elements with atomic numbers 10, 18, and 36 have very small, or zero, electron affinities. Why does an increase of one unit for the atomic number result in such a drastic change in electron affinity?

WHY?

Chemical bonds form when a system of positively-charged nuclei and negatively-charged electrons shift from a higher-energy state to a lower-energy one. This electrical force between pairs of atoms is called a *covalent chemical bond*, because it involves the sharing of valence electrons. The characteristics of chemical bonds lie at the heart of chemistry, and an understanding of these bonds is key to understanding and using chemistry itself.

LEARNING OBJECTIVES

- Identify how the energy of two atoms depends on their separation
- Visualize the energy and force associated with the interaction of charged particles

SUCCESS CRITERIA

- Correctly describe how the energy changes as two atoms approach each other
- Accurately interpret a graph of energy vs. the distance between two atoms to find the bond length and bond energy (which is also called the *bond strength*)

MODEL   THE INTERACTION OF TWO HYDROGEN ATOMS

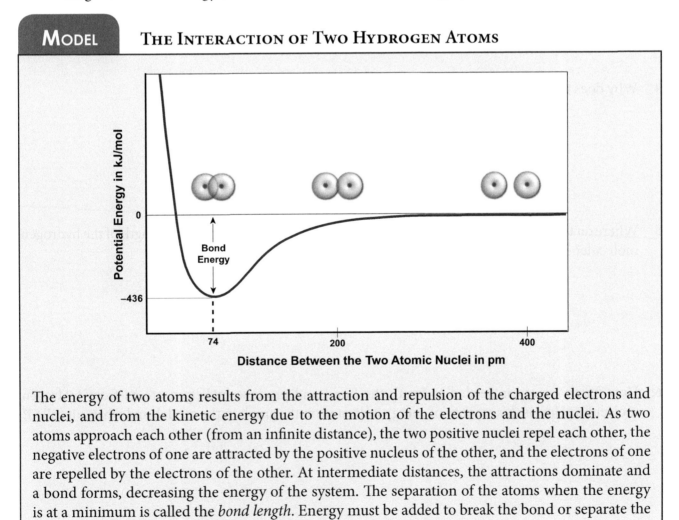

The energy of two atoms results from the attraction and repulsion of the charged electrons and nuclei, and from the kinetic energy due to the motion of the electrons and the nuclei. As two atoms approach each other (from an infinite distance), the two positive nuclei repel each other, the negative electrons of one are attracted by the positive nucleus of the other, and the electrons of one are repelled by the electrons of the other. At intermediate distances, the attractions dominate and a bond forms, decreasing the energy of the system. The separation of the atoms when the energy is at a minimum is called the *bond length*. Energy must be added to break the bond or separate the atoms; this energy is called the *bond energy*.

With contributions by Vicky Minderhout and colleagues at Seattle University

KEY QUESTIONS

1. What quantity is plotted on each axis in the graph shown in the model?

2. Based on the model, how far apart are the two hydrogen atoms when their potential energy is defined as zero?

3. Why does the energy decrease as the two atoms come together? Respond using your own words.

4. Why does the energy increase as the two atoms come very close together?

5. Where on the abscissa in the graph is the distance corresponding to the bond length of the hydrogen molecule? Explain why you selected that particular value.

6. How much energy is needed to break the bond and produce hydrogen atoms from hydrogen molecules? How is this amount of energy indicated on the graph in the model, and what is it called?

1. Construct a diagram showing all the particles that make up a H_2 molecule, using arrows to show all the forces between them. Label each force as repulsive (R) or attractive (A).

Why?

The Lewis model of molecular electronic structure describes how atoms bond with each other to form molecules. It determines the number of bonds formed between pairs of atoms in a molecule and the number of electrons that exist as lone or nonbonding pairs. This information makes it possible for you to predict the geometry of molecules using Valence Shell Electron Pair Repulsion (VSEPR) Theory (e.g., CO_2 is linear but SO_2 and H_2O are bent) and to predict relative bond strengths and lengths.

Learning Objective

- Understand how to draw Lewis structures and use them to predict molecular properties

Success Criteria

- Correctly write Lewis symbols for atoms and monatomic ions
- Construct realistic Lewis structures
- Identify relative bond strengths and lengths from Lewis structures

Prerequisites

03-2 *Representing Molecular Structures*

07-4 *Multi-electron Atoms, the Aufbau Principle, and the Periodic Table*

Information

Molecules exist because they are more stable than separated atoms. By "more stable," we mean that they have lower energy, in the same way that a skateboarder at the bottom of a hill has less energy or is more stable than one at the top. The physical chemist G. N. Lewis recognized, from the very low chemical reactivity of the noble gases, that a configuration with eight electrons in a shell produces a very stable atom. He therefore proposed that molecules form so that atoms can transfer or share electrons and produce this very stable octet structure.

Lewis structures are used to model how electrons are arranged to produce these stable eight-electron configurations. In these diagrams, dots are used to represent nonbonding electrons; a line between atoms represents a single covalent bond formed by a pair of electrons; ions may be explicitly identified with a charge number.

The bonds show how the atoms in a molecule are connected to each other. A Lewis diagram does not show bond lengths, bond angles, the arrangement of atoms in three-dimensional space, or the actual charges on atoms. Some molecules require more than one Lewis structure to describe them. These multiple structures are called *resonance structures*.

In some situations, atoms in Period 2 can have fewer than 8 electrons, and atoms in Periods 3 and higher can have more than 8 electrons. Some atoms (e.g., C, N, O, and S) form double bonds, which are represented by double lines. Some atoms (e.g., C and N) form triple bonds, which are represented by triple lines. A double bond is stronger and shorter than single bond, and a triple bond is the strongest and shortest of the three.

How does one determine and draw a Lewis structure? First determine whether the molecule is ionic or covalent. If it is ionic, draw each ion separately. For covalent molecules and polyatomic ions, follow the methodology given in **Model 2**.

Only valence electrons are illustrated in a Lewis Model, as they are the electrons primarily involved in covalent bonding. In ionic bonding, the loss of valence electrons to form cations or the addition of electrons to complete a set of **p** orbitals are also important.

The valence electrons are those in the **s** and **p** subshells with the highest principle quantum number, **n**. The valence electrons are shown as dots above, below, to the right and to the left of an atomic symbol. The order in which the dots are placed generally does not matter, but dots are dispersed singly before they are paired.

Atomic Symbol	Electron Configuration	Number of Valence Electrons	Lewis Symbol
H	$1s^1$	1	H·
He	$1s^2$	2	·He·
Li	$1s^2 2s^1$ or $[He]2s^1$	1	Li·
Be	$[He]2s^2$	2	·Be·
B	$[He]2s^2 2p^1$	3	·Ḃ·
C	$[He]2s^2 2p^2$	4	·Ċ·
N	$[He]2s^2 2p^3$	5	·Ṅ·
O	$[He]2s^2 2p^4$	6	·Ö·
F	$[He]2s^2 2p^5$	7	:F̈·
Ne	$[He]2s^2 2p^6$	8	:Ne:

Ion Symbol	Electron Configuration	Number of Valence Electrons	Lewis Symbol
H^+	$1s^0$	0	$[H]^+$
Li^+	$[He]$	0	$[Li]^+$
N^{3-}	$[He]2s^2 2p^6$	8	$[·Ṅ·]^{3-}$
O^{2-}	$[He]2s^2 2p^6$	8	$[·Ö·]^{2-}$
F^-	$[He]2s^2 2p^6$	8	$[:F̈·]^-$

KEY QUESTIONS

1. How can you determine how many valence electrons an atom has?

2. How can you determine how many valence electrons an ion has?

3. a. How many valence electrons does H have? How many more does it need to fill its first shell?

    b. How many valence electrons does Cl have? How many more does it need to achieve a noble gas configuration?

MODEL 2 **METHODOLOGY FOR CONSTRUCTING LEWIS STRUCTURES**

HCl – A Simple Example

Methodology		Example: Single Bond
Step 1:	Count the valence electrons from all the atoms. Add electrons for negative ions, subtract electrons for positive ions. The number of valence electrons can be determined from the atom's position in the Periodic Table.	Consider hydrochloric acid: H has 1 valence electron, and Cl has 7, for a total of 8.
Step 2:	Assemble the bonding framework. Decide which atoms are connected to each other, and use a pair of electrons, represented by a line, to form a bond between each pair of atoms that are bonded together.	**H—Cl**
Step 3:	Arrange the remaining electrons so that each atom has 8 electrons around it (the octet rule). If necessary, place additional pairs of electrons (represented by a line) between the atoms to form additional bonds.	**H—C̈l:**
Step 4:	Check for exceptions to the octet rule. For H, only 2 electrons are needed (the duet rule). Be and B are often electron deficient and may have only 4 or 6 electrons. Atoms in the second period cannot have more than 8 electrons. Elements in the third period and higher sometimes have more than 8 electrons.	Cl satisfies the octet rule. H satisfies the duet rule.

Methodology	Example: Single Bond

Step 5: Determine the formal charges (FC) on the atoms.
 	
FC = number of atomic valence electrons
 	
    – number of lone pair electrons
 	
    – 0.5(number of shared electrons).
 	
Evaluate whether the formal charges (FC) on the atoms are reasonable. The structure is reasonable if the charges are zero or small, and the negative charges reside on the most electronegative atoms. Other structures with larger formal charges are higher energy and thus more unstable; these arrangements are not representative of real-world molecular structures..
	$FC(H) = 1 - 0 - 0.5\,(2) = 0$
	
	$FC(Cl) = 7 - 6 - 0.5\,(2) = 0$
	
	Reasonable, because
	FC is zero.

Step 6: Draw resonance structures.	none in this case

4. From a Lewis perspective, why do hydrogen and chlorine combine to form the HCl molecule?

5. From a Lewis perspective, why do lithium and fluorine combine to give an ionic compound?

6. What are the similarities and differences in the Lewis representations of HCl and LiF?

7. How is formal charge determined?

8. How is formal charge used to identify unreasonable Lewis structures?

EXERCISES

1. How many valence electrons do H and F each have? Draw the Lewis structure for HF.

2. How many valence electrons does C have? Determine how many more electrons it needs to form a stable molecule, and draw the Lewis structure for a molecule composed of C and H.

3. How many valence electrons does S have? Determine how many more electrons S needs to form a molecule, and draw the Lewis structure for a molecule composed of S and H.

4. Draw Lewis structures for H_2O, NH_3, PCl_3, C_2H_6, NaCl, and MgO.

Table 1

CO_2 – A Need for Double Bonds

	Methodology	Example: Multiple Bonds
Step 1:	Add up the number of valence electrons from all the atoms.	For carbon dioxide, C has 4 valence electrons, and O has 6 valence electrons, for a total of 16 valence electrons.
Step 2:	Assemble the bonding framework.	**O — C — O**
Step 3:	Arrange the remaining electrons so that each atom has 8 electrons around it (the octet rule). If necessary, place additional pairs of electrons between the atoms to form additional bonds.	$\overset{..}{\underset{..}{O}} = C = \overset{..}{\underset{..}{O}}$
Step 4:	Check for exceptions to the octet rule.	No exceptions present.
Step 5:	Evaluate whether the formal charges on the atoms are reasonable.	FC(C) = 4 – 0 – 0.5 (8) = 0 FC(O) = 6 – 4 – 0.5 (4) = 0 Reasonable because FC is zero.
Step 6:	Draw resonance structures.	none in this case

KEY QUESTIONS

9. How many valence electrons are there in a carbon dioxide molecule?

10. Why is it sometimes necessary when constructing Lewis structures to put double or even triple bonds between atoms?

EXERCISES

5. Draw the Lewis structures for O_2, N_2, C_2H_4, and C_2H_2.

ICl_4^- – Exception to the Octet Rule

	Methodology	Example: Third Period Element
Step 1:	Add up the number of valence electrons from all the atoms, and, because it is an anion with a charge of –1, add one.	ICl_4^-, has 36 valence electrons (5 × 7 + 1).
Step 2:	Assemble the bonding framework.	
Step 3:	Arrange the remaining electrons so that each atom has 8 valence electrons.	
Step 4:	Check for exceptions to the octet rule. After satisfying the octet rule for each atom, 4 electrons remain. These are placed as nonbonding electrons on the fifth period element, iodine.	
Step 5:	Evaluate whether the formal charges on the atoms are reasonable.	FC(I) = 7 – 8 = –1
Note:	*The negative formal charge is on I, which is less electronegative than Cl. This result is correct and is an exception to the general rule given earlier.*	FC(Cl) = 7 – 7 = 0
Step 6:	Draw resonance structures.	none in this case

KEY QUESTIONS

11. How does the example of ICl_4^- differ from the examples of HCl and CO_2?

6. Draw the Lewis structure for PCl_5.

7. Draw the structure for beryllium chloride, $BeCl_2$. Use formal charges to predict the best structure (most stable configuration).

8. Table 2 and Exercises 6 and 7 are examples of *expanded octet* and *electron deficient* Lewis structure exceptions. Based on your understanding of the periodic table and electron configurations, which groups of elements would you expect to adhere to the octet rule? Where would you expect to find expanded octets? What about electron deficiency?

9. Based on your experience with Lewis Structures so far, complete the following table.

| Element | For Zero Formal Charge: Number of... | | Expected Bonding Patterns for Zero Formal Charge |
	Bonds	Lone Pairs	
H			
C			
N			

Element	For Zero Formal Charge: Number of...		Expected Bonding Patterns for Zero Formal Charge
	Bonds	Lone Pairs	
O			
F			

Table 3

NO_2^- – The Need for Multiple Lewis Structures

	Methodology	Example: Resonance
Step 1:	Add up the number of valence electrons from all the atoms, and, because the molecule has a charge of –1, add one.	NO_2^- has 18 valence electrons $(5 + (2 \times 6) + 1)$.
Step 2:	Assemble the bonding framework.	O — N — O
Step 3:	Arrange the remaining electrons so that each atom has 8 electrons around it.	$\left[\ddot{\underset{\cdot\cdot}{O}} - \ddot{N} = \ddot{\underset{\cdot\cdot}{O}} \right]^-$
Step 4:	Check for exceptions to the octet rule.	none
Step 5:	Evaluate whether the formal charges on the atoms are reasonable.	FC(N) = 5 – 5 = 0 FC(O1) = 6 – 7 = –1 FC(O2) = 6 – 6 = 0
Step 6:	In Step 3 the double bond could just as well have been drawn on the left side of the nitrogen. This second structure is called a resonance structure. In fact, experimental measurements show that both N-O bonds are the same with a bond energy and a bond length that are intermediate between a typical N-O single bond and a typical N=O double bond. Therefore, the second structure is needed to adequately describe the bonding. Draw the resonance structure to show that both bonds are equivalent.	$\left[\ddot{\underset{\cdot\cdot}{O}} = \ddot{N} - \ddot{\underset{\cdot\cdot}{O}} \colon \right]^-$

KEY QUESTIONS

12. Why is it not possible to describe NO_2^- with a single Lewis structure?

10. Draw the Lewis structure for ozone, O₃. Include the formal charge in parentheses above each atom if the formal charge differs from zero, e.g., (+1) or (−1).

MODEL 3   BOND STRENGTHS AND LENGTHS

Bond	Average Dissociation Energy in kJ/mole	Average Length in pm
C–C	345	154
C=C	615	133
C≡C	835	120

KEY QUESTIONS

13. The energy it takes to dissociate or break a bond is a measure of bond strength. Considering the data in **Model 3**, how does the Lewis structure help you identify the strongest bonds in a molecule?

14. How does the Lewis structure help you identify the shortest bonds in a molecule? (Consider **Model 3**.)

EXERCISE

11. Identify which C–O bond in acetic acid is the shortest and strongest.

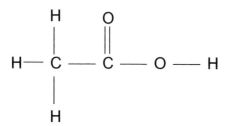

MODEL 4	**WHICH LEWIS STRUCTURE IS BETTER?**

The two structures below both describe the thiocyanate ion. Which one has a lower energy and is therefore a better description of the lowest energy state of this ion?

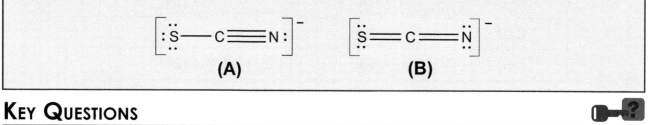

KEY QUESTIONS

15. What are the formal charges on the atoms in the thiocyanate ion in **Model 4**? Write the formal charge in parentheses above each atom in these structures.

16. Based on the criteria developed in Key Question 15, which structure for thiocyanate (in **Model 4**) has the lower energy and is therefore the better description of the lowest energy state of this ion?

12. Taking into account the formal charges, why is $O = C = O$ a better representation of carbon dioxide than $O - C \equiv O$?

INFORMATION

In assembling the bonding framework of a molecule, you may find it difficult to identify which atoms are bonded to each other. Sometimes this information cannot be deduced from first principles: you simply need to know the molecular structure. Here are some guidelines that are often helpful.

- It is useful to think in terms of *outer atoms* and *inner atoms*. An *outer atom* bonds to only one other atom, while an *inner atom* bonds to more than one other atom.

- Hydrogen atoms are always outer atoms because they can form only one bond

- Outer atoms other than hydrogen are usually the ones with the highest electronegativities

- The bonding framework is often indicated by the order in which the atoms are written in a molecular formula. For example, in OCS the carbon atom is the inner atom.

- Parentheses are often used in a molecular formula to indicate the bonding framework. For example in $(CH_3)_2CO$, the hydrogen atoms are bonded to carbon to form two methyl radicals, and oxygen is an outer atom bonded to an inner carbon atom.

- Multiple atoms of the same element are usually outer atoms around a single atom of another element. For example, in PF_6, phosphorus is the inner atom, surrounded by six fluorine atoms.

- Sometimes it is helpful to know the chemical properties of a molecular formula. For example, in HNO_3, the hydrogen atom might be bonded to the nitrogen or to an oxygen. If you know that nitric acid is an oxyacid, you can conclude that the hydrogen is bonded to an oxygen atom.

- Finally, the most likely structure is the one which has the most reasonable formal charges. By "reasonable," we mean that the formal charges should be small or zero, and the negative charges should be located on the most electronegative atoms.

12. Draw Lewis structures for the following molecules. A Lewis structure includes the formal charge on each atom if it differs from zero and any resonance structures that are significant. To be significant, a resonance structure must not increase the formal charge on any of the atoms. Indicate the formal charge on an atom by writing it in parentheses next to the atom, e.g., F (–1).

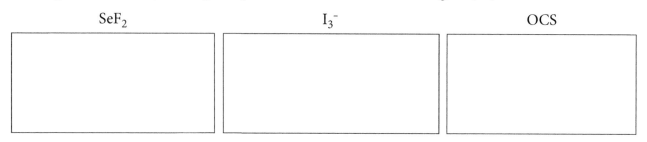

SeF$_2$ I$_3^-$ OCS

HNO$_3$ SiCl$_4$

13. Use Lewis structures to arrange the following compounds in order of increasing carbon-carbon bond **strength**. Explain.

C$_2$H$_6$ (ethane)   C$_2$H$_4$ (ethylene)   C$_2$H$_2$ (acetylene)

14. Use Lewis structures to arrange the following compounds in order of increasing carbon-carbon bond **length**. Explain.

C$_2$H$_6$ (ethane)   C$_2$H$_4$ (ethylene)   C$_2$H$_2$ (acetylene)

15. Explain which N–N bond is more stable: the one in nitrogen (N_2) or the one in hydrazine (N_2H_4).

16. Explain which C–O bond is shorter: the one in methanol (CH_3OH) or the one in formaldehyde (CH_2O).

PROBLEMS

1. Use Lewis structures to identify which of the following compounds you would most likely be successful in synthesizing: SiF_4, OF_4, SF_6, or OF_6. Explain.

2. Two Lewis structures are needed to describe the bonding in formamide, $HCONH_2$. Draw these two resonance structures. One should have no formal charges on the atoms, and the other will have a formal charge of +1 on N and −1 on O. Of the structures you have drawn, which do you expect to be the better description of formamide?

3. For the two resonance structures of formamide in Problem 2, explain why each of the following statements is either correct or incorrect:

   a. The molecule is **not** oscillating back and forth between the two structures; the molecule is, instead, an average or superposition of the two structures.

   b. The expected CO bond length is between that of a normal CO double bond and that of a normal CO single bond.

   c. The nitrogen in the molecule has a lower electron density associated with it than a free nitrogen atom.

   d. Both resonance structures have the same energy.

REFLECTION

Develop a checklist that you can use to ensure that any Lewis structure you have drawn is correct.

WHY?

Molecules adopt a shape that minimizes their stored energy. In many cases, it is possible to predict the geometry of a molecule simply by considering the repulsive energy of its electron pairs. You can use this valence shell electron pair repulsion model (VSEPR) to predict the three-dimensional shapes of molecules. In a later activity, molecular shape will be used to determine whether or not a molecule is polar, which is important in understanding inter-molecular forces of attraction. Scientists commonly use this model when they need to predict or estimate the shape of a molecule.

LEARNING OBJECTIVE

- Understand how molecular shape is predicted from the Lewis structure

SUCCESS CRITERION

- Accurately predict molecular shapes

PREREQUISITE

08-2 *Lewis Model of Electronic Structure*

INFORMATION

The terms *Lewis structure, electronic structure, electron arrangement,* and *electron geometry* are used to describe how the bonding and nonbonding electron pairs are positioned in a molecule. The terms *molecular shape, molecular structure,* and *molecular geometry* are used to describe how the atoms are positioned relative to each other in a molecule.

MODEL 1 — METHODOLOGY FOR DETERMINING MOLECULAR GEOMETRIES (SHAPES OR STRUCTURES) FROM THE VSEPR MODEL

Methodology	Example
Step 1: Draw the Lewis electronic structure.	For sulfur dioxide:
Step 2: Count the number of bonds and nonbonding electron pairs around the central atom.	2 double bonds + 1 nonbonding pair = 3. This number is called the *steric number*.

Methodology	Example
Step 3: Molecules take a shape that minimizes their energy. Arrange the bonds and nonbonding electron pairs to maximize their separation, which minimizes the electron-electron repulsion energy.	A steric number of 3 in step 2 means a trigonal planar electronic structure minimizes the energy: 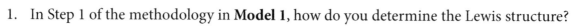
Step 4: Add the atoms in a way that is consistent with how the electrons are shared, and put the nonbonding electron pairs as far apart as possible.	
Step 5: Determine the molecular shape from the position of the atoms.	The atoms are arranged in a nonlinear or bent shape.

KEY QUESTIONS

1. In Step 1 of the methodology in **Model 1**, how do you determine the Lewis structure?

2. Why are bonds and nonbonding electron pairs (aka: *lone pairs*) spaced as far apart as possible in the structure?

3. According to Step 4 in the methodology, if you have two lone pairs and bonds to four atoms around a central atom, would you position the lone pairs at 90° or 180° to each other? Explain.

4. How would you describe the geometrical arrangement of the bonds and lone pairs around the central oxygen in ozone?

5. How would you describe the shape of ozone?

6. Some triatomic molecules are linear. What feature of O_3 explains its bent geometry?

7. What three insights has your team gained about the shape of molecules by examining the model and responding to the key questions?

INFORMATION

Molecular geometries are varied, and depend on the reactions between lone pairs of electrons and bonds between atoms. These geometries are summarized below. Notice the use of the formula AX_nE_m, where A designates the central atom, X_n refers to **n X** atoms bonded to the central atom, and E_m represents **m** lone pairs of electrons on the **A** atom. It is important to learn the geometry names and bond angles.

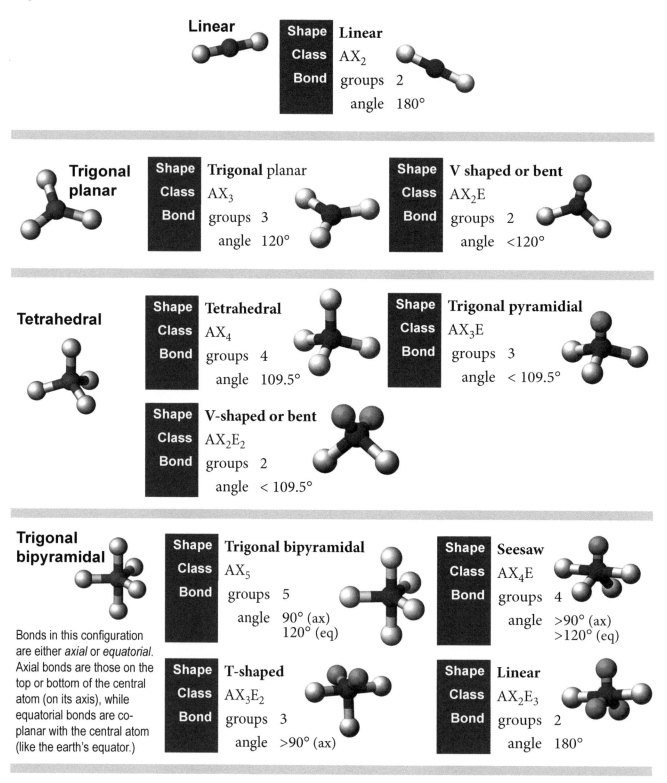

Linear

Shape	Linear
Class	AX_2
Bond groups	2
angle	180°

Trigonal planar

Shape	Trigonal planar
Class	AX_3
Bond groups	3
angle	120°

Shape	V shaped or bent
Class	AX_2E
Bond groups	2
angle	<120°

Tetrahedral

Shape	Tetrahedral
Class	AX_4
Bond groups	4
angle	109.5°

Shape	Trigonal pyramidial
Class	AX_3E
Bond groups	3
angle	< 109.5°

Shape	V-shaped or bent
Class	AX_2E_2
Bond groups	2
angle	< 109.5°

Trigonal bipyramidal

Bonds in this configuration are either *axial* or *equatorial*. Axial bonds are those on the top or bottom of the central atom (on its axis), while equatorial bonds are co-planar with the central atom (like the earth's equator.)

Shape	Trigonal bipyramidal
Class	AX_5
Bond groups	5
angle	90° (ax) 120° (eq)

Shape	Seesaw
Class	AX_4E
Bond groups	4
angle	>90° (ax) >120° (eq)

Shape	T-shaped
Class	AX_3E_2
Bond groups	3
angle	>90° (ax)

Shape	Linear
Class	AX_2E_3
Bond groups	2
angle	180°

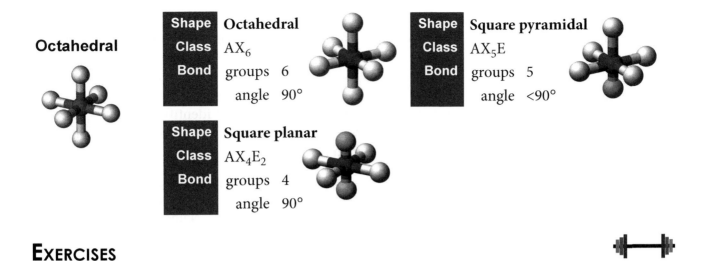

Octahedral

Shape	Octahedral
Class	AX_6
Bond	groups 6
	angle 90°

Shape	Square pyramidal
Class	AX_5E
Bond	groups 5
	angle <90°

Shape	Square planar
Class	AX_4E_2
Bond	groups 4
	angle 90°

EXERCISES

1. Complete the illustrations in the following table, determining the arrangement of bonds and lone pairs that minimizes the atom's energy in each case. Remember: your illustration represents the *Lewis electronic structure* of the molecule. The number of bonds and lone pairs is called the *steric number*.

Table 1

Number of Bonds and Lone Pairs	Lewis Electronic Structure	Illustration of the Electronic Structure
2	linear	
3	trigonal planar	
4	tetrahedral	
5	trigonal bipyramidal	
6	octahedral	

2. All of the possible molecular shapes for atoms arranged around a central atom are shown in the Information section preceding these Exercises. Each of these shapes is exemplified by one molecule from the following list:

$$O_3 \quad I_3^- \quad IF_6^+ \quad SbF_5 \quad COCl_2 \quad SeO_3^{2-} \quad SiF_4 \quad KrF_4 \quad SF_4 \quad ICl_3 \quad BrF_5$$

First, determine the Lewis structure for each molecule in the list and identify the AX_nE_m formula that describes that structure. Use Table 1 and the Information section, as needed, to determine the electron pair geometry and the molecular geometry of each species.

Molecule	Lewis Structure	AX_nE_m formula	Electron Pair Geometry Name	Molecular Geometry Name
O_3				
I_3^-				
$COCl_2$				
KrF_4				
SiF_4				

Molecule	Lewis Structure	AX_nE_m formula	Electron Pair Geometry Name	Molecular Geometry Name
IF_6^+				
SbF_5				
SeO_3^{2-}				
SF_4				
ICl_3				

Molecule	Lewis Structure	AX$_n$E$_m$ formula	Electron Pair Geometry Name	Molecular Geometry Name
BrF$_5$				

3. Based on your results using the VSEPR model, indicate which molecule (from Exercise 2) corresponds with which molecular geometry in Table 3 below.

Table 3

Structure Name/Example	Illustration
a.	
b.	
c.	
d.	
e.	

Structure Name/Example	Illustration
f.	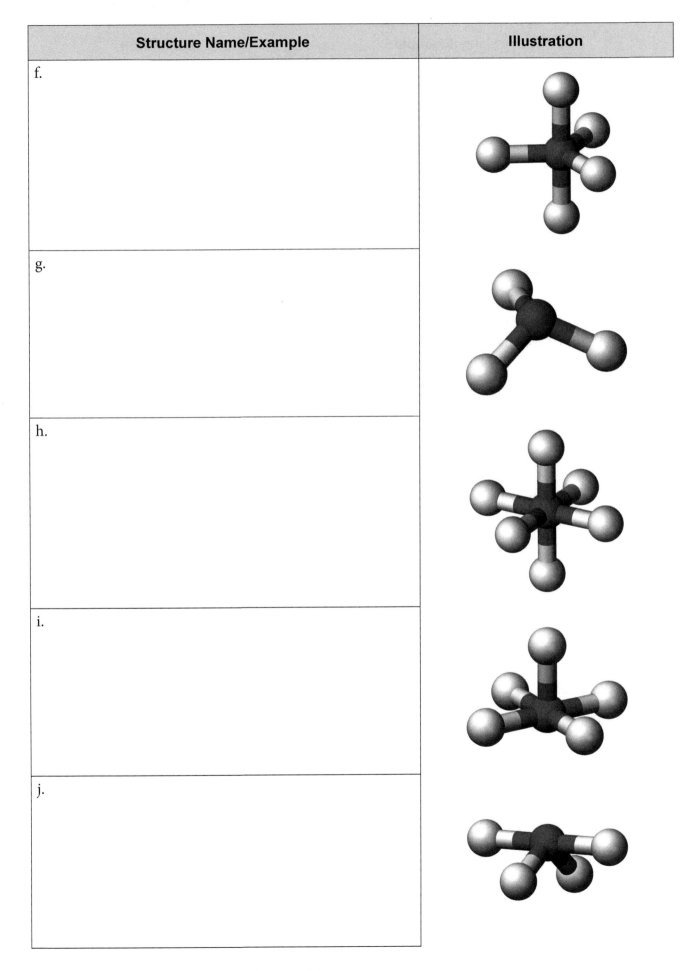
g.	
h.	
i.	
j.	

Structure Name/Example	Illustration
k.	

MODEL 2   PREDICTING DEVIATIONS FROM IDEAL MOLECULAR GEOMETRIES

The repulsive forces between adjacent pairs of bonding and nonbonding electrons are not all equivalent. Lone pairs of electrons repel other pairs of electrons to a greater degree than bonding pairs of electrons, and the result is that nonbonding pairs occupy a greater share of space around a central atom than the ideal geometry would imply. This is also true for double and triple bonds, since the greater electron density gives rise to greater repulsion than a single bond.

Molecule	Lewis Structure	Ideal Bond Angles	Actual Bond Angles	Cause
NH_3	H—N—H with lone pair on N and H below	109.5° H–N–H	Less than 109.5°	Lone pair occupies more space, forcing H atoms closer together
C_2H_4	H₂C=CH₂	120° H–C–H	Less than 120°	Double bond occupies more space, forcing H atoms closer together

KEY QUESTIONS

8. Why is 109.5° ammonia's ideal bond angle?

9. Would you expect the H–O–H bond angle to be greater than or less than ammonia's H–N–H bond angle? Explain using a Lewis structure.

Chapter 8: Covalent Bonding I

1. An article in a journal, *Inorganic Chemistry*, cites both BF_3 and PF_3 as examples of flat or planar molecules with bond angles of 120°. Another article reports the FPF bond angle as 98°. Which report is consistent with the VSEPR model? Explain.

2. Is the molecular geometry of OCS like that of CO_2 or SO_2? Identify which of these molecules are linear and which are bent.

Electronegativity and Bond Polarity

WHY?

Electronegativity is a measure of the ability of an atom in a molecule to attract electrons. The difference in the electronegativities of two atoms profoundly affects the properties of the chemical bond between them and, consequently, has dramatic effects on the physical and chemical properties of materials that contain that bond. You therefore need to be able to identify the origin of ionic bonds and covalent bonds based on electronegativity differences. Furthermore, it is also important to be able to describe the origin of polar covalent bonds—and, ultimately, polar molecules—from atomic electronegativities and molecular geometries (i.e., shapes).

LEARNING OBJECTIVES

- Master using electronegativity to predict bond characteristics
- Identify how bond characteristics affect material properties

SUCCESS CRITERIA

- Correctly organize bonds in order of increasing polarity
- Correctly identify polar molecules

PREREQUISITES

07-5 *Periodic Trends in Atomic Properties*

08-3 *Valence Shell Electron Pair Repulsion Model*

MODEL — ELECTRONEGATIVITY (EN) AND BOND CHARACTERISTICS

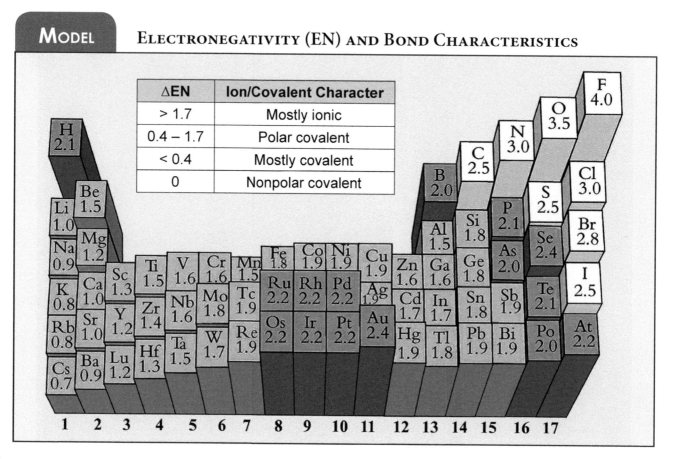

ΔEN	Ion/Covalent Character
> 1.7	Mostly ionic
0.4 – 1.7	Polar covalent
< 0.4	Mostly covalent
0	Nonpolar covalent

1. What is the general trend in electronegativity as you progress across the rows of the Periodic Table?

2. What is the general trend in electronegativity as you progress down the columns of the Periodic Table?

3. What kind of a bond is formed between two atoms with the same electronegativity?

4. What kind of a bond is formed between two atoms that have very different electronegativities?

5. What is the relationship between electronegativity and bond polarity?

EXERCISES

1. Using position in the Periodic Table as the criterion (do not look at a table or chart of electronegativity values), arrange the elements in each of the following groups in order of increasing electronegativity. Use the < symbol in your arrangement.

(F, C, N, O)   _____

(Cl, F, I, Br)   _____

(O, S, F)   _____

(Cs, Ca, Cu, S, Se, Cl)   _____

(As, F, Sb, Cl, S, Se)   _____

2. Perform tasks a) through d) for each of the following groups of bonds:

	Ⓐ	Ⓑ	Ⓒ	Ⓓ
Bonds:	(C–F, Si–F)	(C–F, C–Cl, C–Br)	(N–O, N–F, C–F)	(Cl–Cl, F–Cl, F–Br, Br–Cl)

a. Order the bonds in order of increasing bond polarity. Use the < symbol in your arrangement.

Ⓐ	Ⓑ
Ⓒ	Ⓓ

b. Identify the atom in each bond that carries a slight positive charge. Illustrate this partial positive charge with a δ+ (Greek letter delta) symbol. Indicate the partial negative charge with a δ– symbol.

Ⓐ	Ⓑ
Ⓒ	Ⓓ

c. Use an arrow (⊢––▶) to indicate the direction and extent of electron density shift in each bond, based on electronegativity values, with no arrow representing a nonpolar covalent bond. The arrowhead points to the atom that is more electronegative. These arrows are called *bond dipoles.*

	Ⓐ	Ⓑ	Ⓒ	Ⓓ
Bonds:	(C–F, Si–F)	(C–F, C–Cl, C–Br)	(N–O, N–F, C–F)	(Cl–Cl, F–Cl, F–Br, Br–Cl)

Ⓐ	Ⓑ
Ⓒ	Ⓓ

d. Classify each bond as mostly ionic, polar covalent, mostly covalent, or nonpolar covalent based on electronegativity values.

Bonds:

A	**B**	**C**	**D**
(C–F, Si–F)	(C–F, C–Cl, C–Br)	(N–O, N–F, C–F)	(Cl–Cl, F–Cl, F–Br, Br–Cl)

A

	C–F	Si–F
mostly ionic	O	O
polar covalent	O	O
mostly covalent	O	O
nonpolar covalent	O	O

B

	C–F	C–Cl	C–Br
mostly ionic	O	O	O
polar covalent	O	O	O
mostly covalent	O	O	O
nonpolar covalent	O	O	O

C

	N–O	N–F	C–F
mostly ionic	O	O	O
polar covalent	O	O	O
mostly covalent	O	O	O
nonpolar covalent	O	O	O

D

	Cl–Cl	F–Cl	F–Br	Br–Cl
mostly ionic	O	O	O	O
polar covalent	O	O	O	O
mostly covalent	O	O	O	O
nonpolar covalent	O	O	O	O

3. Identify whether metals and nonmetals always form mostly ionic bonds. Can you find metal-nonmetal pairs that produce mostly covalent or polar covalent bonds?

INFORMATION

The overall molecular dipole is determined by adding the individual bond dipoles. Both the magnitudes of the bond dipoles and the geometry of the molecule contribute to this addition process because, if two bond dipoles with the same magnitude are pointing in opposite directions, then they will offset each other, resulting in a nonpolar molecule.

For any AX_n structure; that is, one in which there are no lone pairs and all the X atoms are the same element, the bond dipoles cancel, making the molecule nonpolar. If there are lone pairs, in most cases, the broken symmetry leads to an imbalance in the dipole moments, resulting in a polar molecule. A couple of exceptions arise with the linear AX_2E_3 structure and the square planar AX_4E_2 structure, since both the lone pairs and the X atoms in these molecules are arranged symmetrically.

If the X atoms are not all the same, the molecule will be polar due to differing electronegativity values. An important generalization can be made regarding hydrocarbons (compounds consisting of only carbon and hydrogen, with multiple central carbon atoms). These are generally nonpolar molecules due to the absence of lone pairs and the symmetry at each carbon, even for molecules with many carbon atoms in C–C bonds.

EXERCISES

4. Determine which of the following molecules are polar and illustrate the directions of the molecular dipoles. Note that lone pairs are not shown in the structures. The terms *cis* and *trans* refer to the placement of similar atoms in cases where both X and Y atoms are present around the central A atom. The term *cis* indicates the similar atoms are next to each other, while the term *trans* indicates that they are across from each other.

Molecule	Structure	Polar or nonpolar?
CO_2	O=C=O	
H_2O		
NH_3		
NF_3		
NO_3^-		
$CHCl_3$		
BF_3		
CH_2F_2		

Molecule	Structure	Polar or nonpolar?
cis- PCl$_3$F$_2$	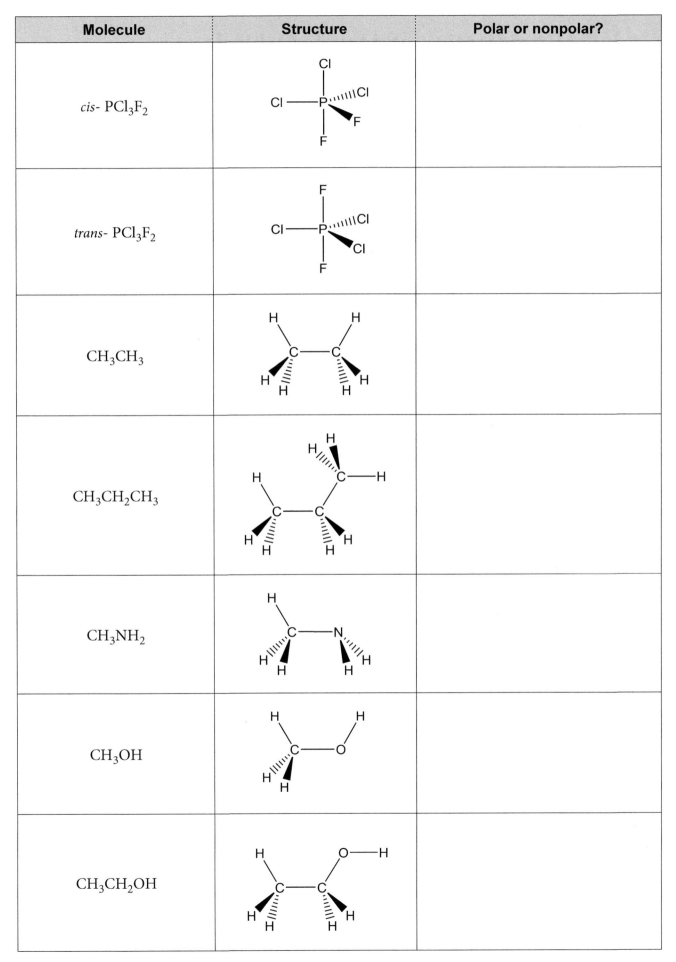	
trans- PCl$_3$F$_2$		
CH$_3$CH$_3$		
CH$_3$CH$_2$CH$_3$		
CH$_3$NH$_2$		
CH$_3$OH		
CH$_3$CH$_2$OH		

ACTIVITY 09-1 Hybridization of Atomic Orbitals

WHY?

In the same way that the locations of electrons within atoms can be described by atomic orbitals, covalent bonds can be described with mathematical functions called *bonding orbitals*. While atomic orbitals can also be written in other ways, chemists like to use atomic orbitals because this approach appeals to their intuition about how electrons are arranged around the atoms in a molecule, and about how molecules are produced from atoms. However, the reality of atomic orbitals differs from the ideal. Atomic p-orbitals are oriented at angles of 90° to one another, but bond angles in molecules are rarely a perfect 90°. To account for bond angles that differ from 90°, atomic orbitals can be combined to form *hybrid orbitals*, in a process called, appropriately, *hybridization*. You can use the geometries predicted by the VSEPR model to determine these hybrid orbitals.

LEARNING OBJECTIVE

- Master the geometric properties of hybrid orbitals

SUCCESS CRITERIA

- Correctly identify hybrid orbitals from molecular geometry
- Correctly distribute valence electrons in hybrid orbitals
- Use hybrid orbitals to explain single bonds and lone pairs

PREREQUISITES

08-1 *The Chemical Bond*

08-2 *Lewis Model of Electronic Structure*

08-3 *Valence Shell Electron Pair Repulsion Model*

INFORMATION

As seen in the formation of H_2 in Activity 8-1, the overlap of atomic orbitals of similar energies results in a covalent bond if each atom contributes one electron. However, the actual geometries of most molecules require that the atomic orbitals become hybridized, or blended, creating hybrid atomic orbitals that overlap with the orbitals of the bonded atoms. The Lewis structure and electron pair geometry of the molecule determined by the VSEPR model indicates the number of hybrid orbitals required and, therefore, the number of atomic orbitals involved in the hybridization. When orbitals hybridize, the resulting hybrid orbitals all have the same energy, and valence electrons are distributed across these orbitals.

The orbital hybridization corresponding to each electron pair geometry is summarized in the following table:

Electron Pair Geometry	Formulas	Steric Number	Hybridization
Linear	AX_2, AXE	2	sp
Trigonal Planar	AX_3, AX_2E, AXE_2	3	sp^2
Tetrahedral	AX_4, AX_3E, AX_2E_2, AXE_3	4	sp^3
Trigonal Bipyramidal	AX_5, AX_4E, AX_3E_2, AX_2E_3	5	sp^3d
Octahedral	AX_6, AX_5E, AX_4E_2	6	sp^3d^2

MODEL 1 — HYBRID ORBITALS IN METHANE, AMMONIA, AND WATER

For methane, CH_4, ammonia, NH_3, and water, H_2O, the electron pair geometries are all tetrahedral due to the single lone pair in ammonia and two lone pairs in water.

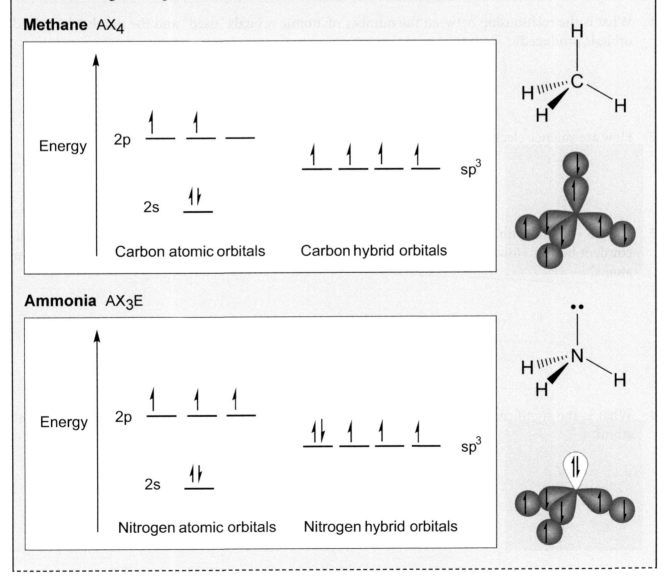

Water AX_3E_2

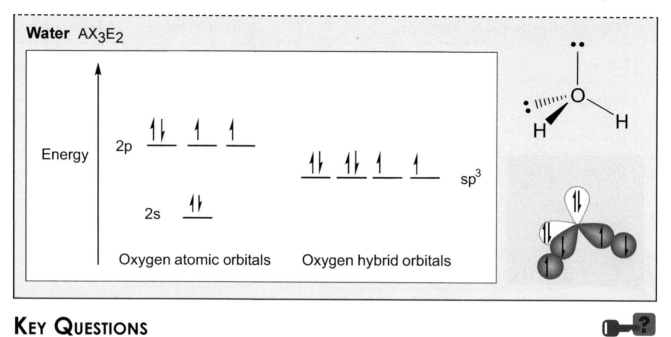

Energy

2p

2s

Oxygen atomic orbitals   Oxygen hybrid orbitals

sp^3

KEY QUESTIONS

1. What is the relationship between the number of atomic orbitals "used" and the number of hybrid orbitals produced?

2. How are valence electrons distributed across hybrid orbitals?

3. Hybrid orbitals with a single electron are able to overlap with hydrogen atoms to form single covalent bonds. How many electrons will be donated to each unfilled hybrid orbital by hydrogen atoms?

4. What is the significance of the hybrid orbitals with a **pair** of valence electrons from the central atom?

Both the ammonium ion, NH_4^+, and the hydronium ion, H_3O^+, have a tetrahedral molecular geometry, as in ammonia and water. However, the formal charges on the nitrogen atom and the oxygen atom in these molecules are both +1. The number of valence electrons on the ion is therefore decreased by 1 in each case.

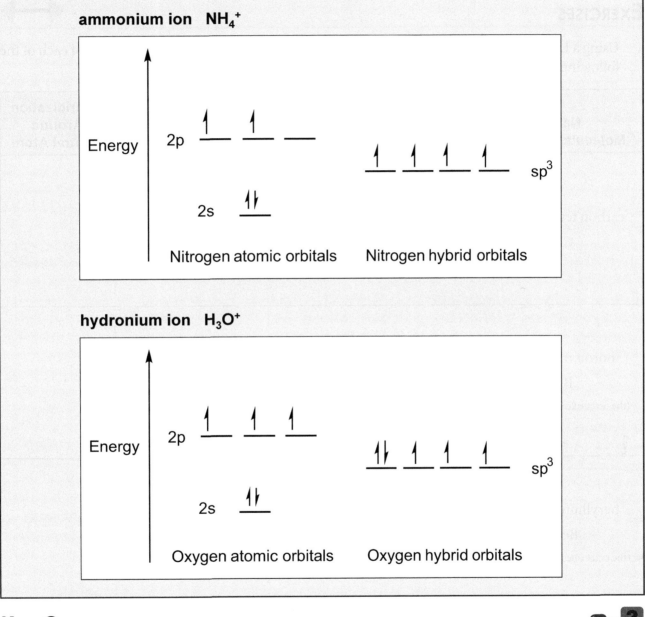

KEY QUESTIONS

5. Why is it necessary to decrease the number of valence electrons held by nitrogen and oxygen atoms, respectively, in the ammonium and hydronium ions?

6. What is the result of decreasing the number of valence electrons on the nitrogen and oxygen atoms?

EXERCISES

1. Using a Lewis structure and the VSEPR model, determine the electron pair geometry of each of the following molecules. Identify the hybrid orbitals of the central atom in each.

Name Molecular Formula	Lewis Structure	Molecular Shape, Expected Bond Angles	Hybridization Around Central Atom
carbon tetrafluoride CF_4			
boron trifluoride BF_3 (the octet rule is not followed)			
beryllium hydride BeH_2 (the octet rule is not followed)			
phosphorus pentachloride PCl_5 (the octet rule is not followed)			

ACTIVITY

09-2 Valence Bond Model for Covalent Bonds

WHY?

Covalent bonds are formed by atoms sharing pairs of electrons. We use Lewis structures to denote the **locations** of these bonds. However, Lewis structures do not provide any information about the **orbitals** occupied by either the bonding or the lone pair electrons. The valence bond model serves this purpose instead; it describes a covalent bond formed from the overlap of orbitals belonging to each atom. The overlap between the two atoms builds up electron density, which attracts the positively charged atomic nuclei and holds the atoms together. Matching orbital geometries are necessary for two atoms to form a bonding orbital, and different geometries result in different kinds of orbitals. The valence bond model accounts for the fact that the orbitals of different bonds do not all overlap the same amount; the greater the overlap, the stronger the bond becomes.

LEARNING OBJECTIVE

- Learn to characterize sigma (σ) and pi (π) bonds in terms of overlapping atomic orbitals

SUCCESS CRITERIA

- Correctly describe sigma (σ) and pi (π) bonds, including descriptions of overlapping atomic orbitals
- Correctly describe single, double, and triple bonds in terms of sigma (σ) and pi (π) bonds

PREREQUISITES

07-3   *Quantum Numbers for Electrons in Hydrogen Atoms*

08-2   *Lewis Model of Electronic Structure*

09-1   *Hybridization of Atomic Orbitals*

MODEL 1 ATOMIC ORBITAL OVERLAP IN THE FORMATION OF SIGMA (σ) BONDS

When two atomic orbitals from adjacent atoms, each occupied with one electron, overlap or share space in the region directly between the nuclei, the resulting bonding orbital is called a sigma (σ) bond. This is the case for the formation of the H_2 molecule. Remember: $s + s = \sigma$. The two electrons form a pair with opposite m_s values just as they do in atomic orbitals and fill the bonding orbital.

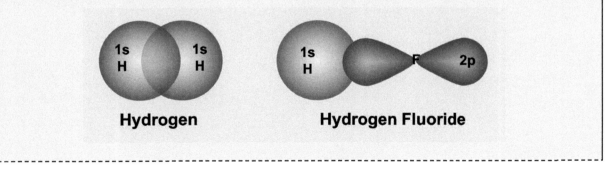

Hydrogen   Hydrogen Fluoride

For *any* pair of adjacent atoms with the direct inter-nuclear overlap that is seen in H_2, if each of the two atomic orbitals is occupied with one electron, sigma bonding will result. In the HF Lewis structure, four electron pairs on the fluorine atom indicate sp^3 hybridization.

$$H\!\!-\!\!\overset{..}{\underset{..}{F}}:$$

With seven valence electrons, three of these sp^3 hybrid orbitals are fully occupied with spin-paired electrons (these make up the three lone pairs.) The one half-occupied orbital faces the hydrogen atom and forms a σ bond.

KEY QUESTIONS

1. Based on the model, what types of atomic orbitals are overlapping in each case?

    a. hydrogen:

    b. hydrogen fluoride:

2. What types of atomic orbitals overlap in the formation of C–H bonds in methane, CH_4? How many electrons are in each atomic orbital? What kind of bonds are formed?

3. What types of atomic orbitals overlap in the formation of F_2? What type of bond is formed? Explain.

In the Lewis structure of the oxygen molecule, O_2, there are two lone pairs on each oxygen atom and a double bond between the atoms as shown in the figure.

:O═O:

If we consider the steric number at each of each oxygen atom to be three, due to the two lone pairs and one bonded atom (even though it is a double bond), then we would determine that both atoms have a trigonal planar geometry, with sp^2 hybridization. The formation of these hybrid orbitals and the distribution of the six valence electrons shows that there are more electrons available than would be necessary to form a sigma bond with two leftover lone pairs. Since the hybridization uses only two of the p orbitals, there is one unhybridized p orbital left unchanged. The sixth electron remains in that orbital, and is available for bond formation. However, the geometry of this orbital, with two lobes of electron density, does not allow sigma bond formation with the adjacent oxygen atom. Since each oxygen atom is identical, and each has a half-filled p orbital, another type of bond, called a *pi bond* (π), is formed. Remember p + p = π. The first bond formed in all molecules is a σ (sigma) bond. If there are additional bonds, they are π bonds (at least for general chemistry).

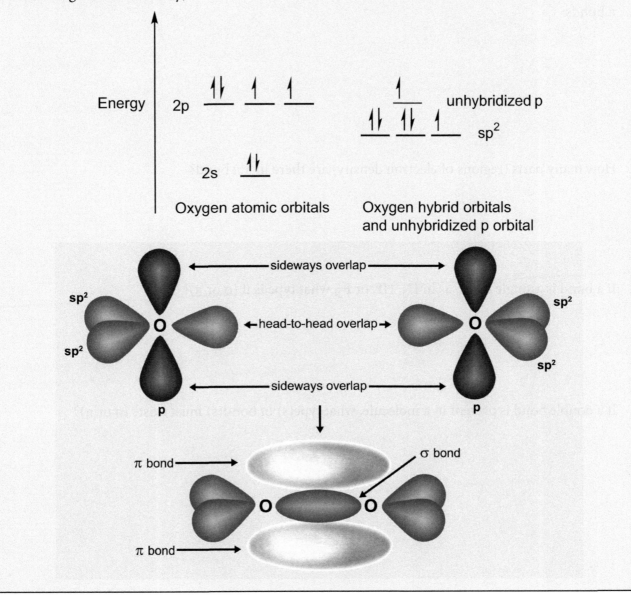

KEY QUESTIONS

4. What is the characteristic geometry of σ and π bonds: *end-to-end* or *side-to-side?*

    σ bonds:

    π bonds:

5. What is the characteristic shape of the following bonds with respect to the line connecting the two nuclei? (This line is also called the *internuclear axis* or the *bond axis*.)

    σ bonds:

    π bonds:

6. How many parts (regions of electron density) are there in a π bond?

7. If a bond is a single bond, as in H_2, HF, or F_2, what type is it (σ or π)?

8. If a double bond is present in a molecule, what type(s) of bond(s) must exist? (σ or π)?

1. Nitrogen, N_2, has a triple bond in its Lewis structure. Sketch the Lewis structure, identify the orbital hybridization on each nitrogen atom, and draw the orbital hybridization energy diagram. Explain the formation of all three bonds, and account for the lone pairs.

2. Carbon monoxide has a triple bond consisting of one σ bond and two π bonds. Draw a diagram to show how one π bond can be formed from the overlapping p_x orbitals of carbon and oxygen, and the other from the overlapping p_y orbitals.

3. Using a Lewis structure, the VSEPR model and the valence bond model, determine the electron pair geometry and molecular geometry of each of the following molecules, a through e. Identify the hybrid orbitals of each atom, and identify the atomic orbitals (including hybrid orbitals) contributing to each bond. Classify each bond as either σ or π.

a.

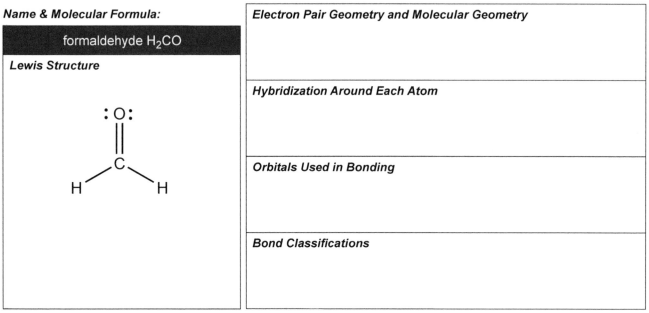

Name & Molecular Formula:	*Electron Pair Geometry and Molecular Geometry*
sulfur dioxide SO$_2$	
Lewis Structure	
	Hybridization Around Each Atom
	Orbitals Used in Bonding
	Bond Classifications

b.

Name & Molecular Formula:	*Electron Pair Geometry and Molecular Geometry*
formaldehyde H$_2$CO	
Lewis Structure	
	Hybridization Around Each Atom
	Orbitals Used in Bonding
	Bond Classifications

c.

Name & Molecular Formula:

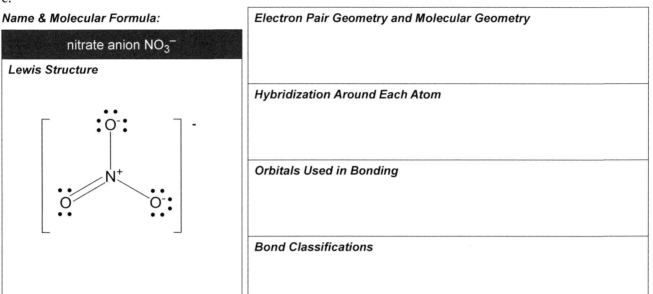

nitrate anion NO_3^-

Lewis Structure

Electron Pair Geometry and Molecular Geometry
Hybridization Around Each Atom
Orbitals Used in Bonding
Bond Classifications

d.

Name & Molecular Formula:

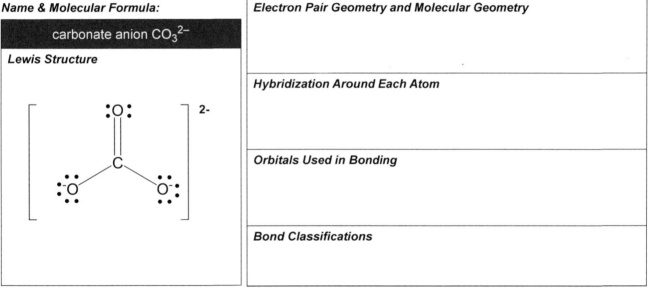

carbonate anion CO_3^{2-}

Lewis Structure

Electron Pair Geometry and Molecular Geometry
Hybridization Around Each Atom
Orbitals Used in Bonding
Bond Classifications

e.

Name & Molecular Formula:

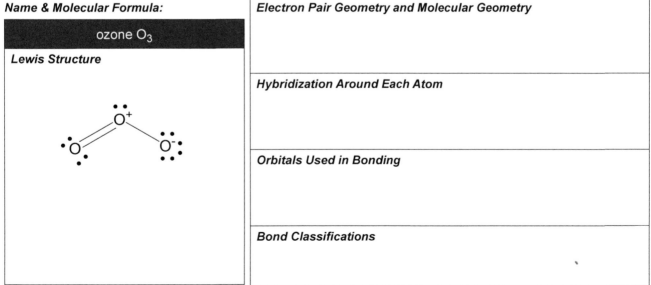

ozone O_3

Lewis Structure

Electron Pair Geometry and Molecular Geometry
Hybridization Around Each Atom
Orbitals Used in Bonding
Bond Classifications

PROBLEMS

1. Given the following structure, number the carbon atoms 1 through 5 starting from the left.

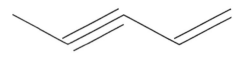

a. Identify the hybrid orbitals on each carbon atom. Be careful; the actual geometry may differ from the way the molecule is drawn. Use the bonding properties of carbon to guide you.

1 _____ 2 _____ 3 _____ 4 _____ 5 _____

b. Determine the local geometry of each carbon atom and predict the bond angles around each.

Carbon Atom	Local Geometry	Bond Angles
1		
2		
3		
4		
5		

2. Identify how the atomic orbitals of N, C, and O are hybridized in formamide. (The formamide structure is a fundamental linking unit in proteins). Explain your reasoning.

WHY?

Just as atomic orbitals describe electrons in atoms, molecular orbitals describe electrons in molecules. Molecular orbitals account for the delocalization of electrons in the molecule and provide an alternative to the Lewis model and Valence Bond model, both of which view electrons as being localized in bonds between pairs of atoms. You can use the molecular orbital model to determine bond order and, consequently, predict the strength, energy, and length of bonds and the magnetic properties of molecules. In this activity, for simplicity's sake, only diatomic molecules will be considered, but molecular orbital theory is applied to larger molecules in more advanced treatments.

LEARNING OBJECTIVES

- Identify how both bonding and anti-bonding molecular orbitals are formed from atomic orbitals
- Understand how electron occupation of molecular orbitals is related to bond order, bond strength, bond energy, bond length, and magnetic properties

SUCCESS CRITERIA

- Construct and identify bonding and anti-bonding molecular orbitals
- Analyze bond order, bond strength, bond energy, bond length, and magnetic properties in terms of the occupation of molecular orbitals

PREREQUISITES

07-5 *Periodic Trends in Atomic Properties*

08-1 *The Chemical Bond*

08-2 *Lewis Model of Electronic Structure*

INFORMATION

Molecular orbitals can be written as sums of atomic orbitals multiplied by a coefficient, which is often +1 or –1. Some of these orbitals are *bonding*, some *nonbonding*, and some are *antibonding*.

A *bonding orbital* has a high electron density between two atomic nuclei. This negative electron charge attracts the positive nuclei and holds them together.

A *nonbonding orbital* positions electrons away from the bonds and does not contribute to bonding; e.g., nonbonded or lone pairs in the Lewis model would be in nonbonding orbitals in the molecular orbital model.

An *antibonding orbital* has little electron density between the two atomic nuclei, allowing the nuclei to repel each other. This repulsive effect opposes the attractive effect of electrons in bonding orbitals, so molecules with equal occupations of bonding and antibonding orbitals are not stable.

Bond order is equal to half the number of bonding electrons minus half the number of antibonding electrons: BO = ½(*n* bonding – *n* antibonding). Bonding electrons attract the atomic nuclei together, and antibonding electrons destabilize this bond. Bond order correlates with bond strength, bond energy, and bond length. The higher the bond order, the stronger the bond; the higher the bond energy, the shorter the bond length. Bond orders predicted by the Lewis model are all integer values (1, 2, 3…), but bond orders determined by molecular orbital theory can be fractions.

MODEL 1 · VALENCE ORBITALS OF DIATOMIC MOLECULES

Notation

Figure 1

atomic orbitals on atom labeled A: 2s(A), $2p_x$(A), $2p_y$(A), $2p_z$(A)

atomic orbitals on atom labeled B: 2s(B), $2p_x$(B), $2p_y$(B), $2p_z$(B)

σ designates a sigma molecular orbital, which has its maximum electron density along a line connecting the two nuclei. This line is called the *internuclear axis*.

π designates a pi molecular orbital, which has its maximum electron density above and below the internuclear axis

$*$ designates an antibonding molecular orbital, which has a region of very low electron density between the two nuclei, due to the presence of a nodal plane

$+/-$ designates the phase of the orbital, which results from constructive or destructive overlap between the orbitals, which are wave-functions

Valence Molecular Orbitals Formed from Atomic Orbitals

Figure 2

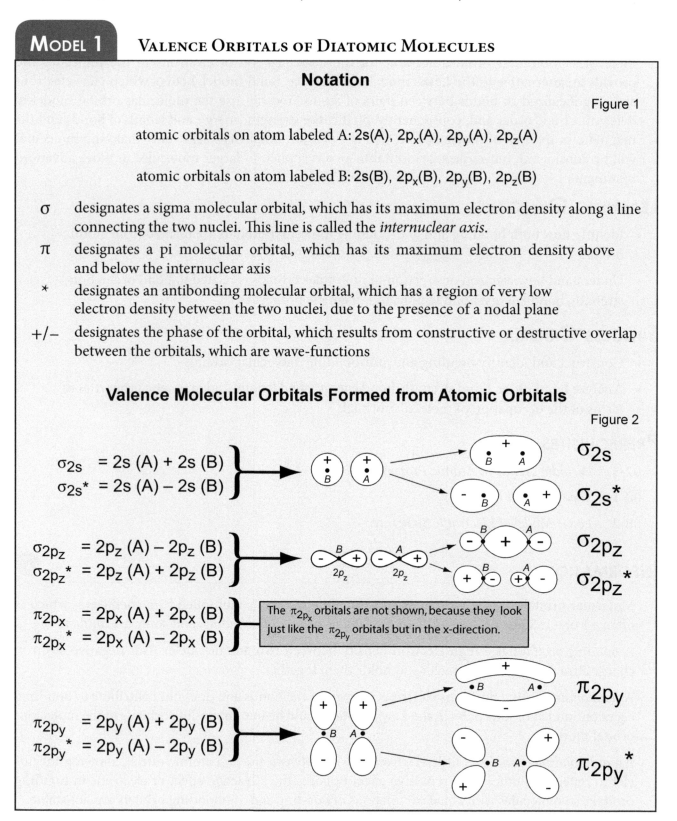

1. How many molecular orbitals are produced by combining two atomic orbitals, as illustrated in **Model 1**?

2. If two **s** atomic orbitals from different atoms are combined, what are the names of the molecular orbitals that are produced? Where are the regions of high and low electron density for these orbitals?

3. If two p_z atomic orbitals from different atoms are combined, what are the names of the molecular orbitals that are produced? Where are the regions of high and low electron density for these orbitals?

4. If two p_y atomic orbitals from different atoms are combined, what are the names of the molecular orbitals that are produced? Where are the regions of high and low electron density for these orbitals?

5. According to **Model 1**, what are the differences between a bonding and an antibonding molecular orbital? (These differences apply to all three pairs of bonding and antibonding orbitals shown in **Model 1**.)

6. In terms of the relationship each has to the internuclear axis, what is the difference between a π and a σ molecular orbital?

7. What are the similarities between a σ_{2s} and a σ_{2pz} molecular orbital?

8. What are the differences between a σ_{2s}^* and a σ_{2pz}^* molecular orbital?

9. In general, antibonding orbitals have higher energy levels than bonding orbitals, due to the increase in the number of nodal planes present. Rank each pair of molecular orbitals in the model according to their relative energies.

EXERCISES

1. Write the π_{2px} and π_{2px}^* molecular orbitals in terms of the $2p_x$ atomic orbitals.

2. Draw diagrams similar to those in **Model 1** to show how the π_{2px} and π_{2px}^* molecular orbitals are formed from the $2p_x$ atomic orbitals.

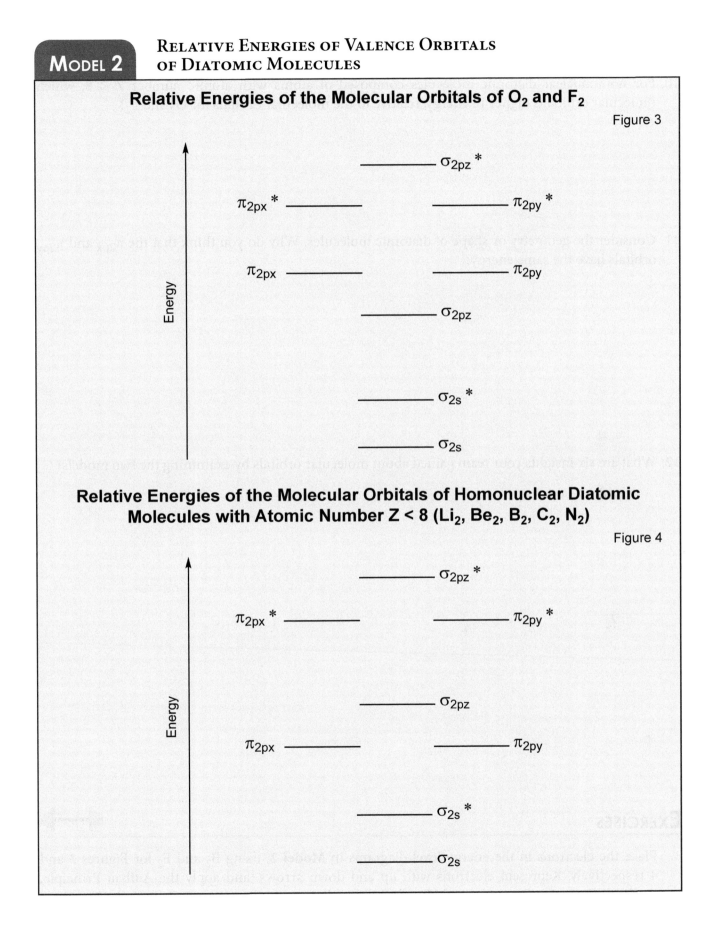

Relative Energies of the Molecular Orbitals of O_2 and F_2

Figure 3

Relative Energies of the Molecular Orbitals of Homonuclear Diatomic Molecules with Atomic Number Z < 8 (Li_2, Be_2, B_2, C_2, N_2)

Figure 4

Key Questions

10. For homonuclear diatomic molecules composed of atoms with atomic number Z < 8, which molecular orbital is higher in energy, relative to the molecular orbitals for O_2 and F_2?

11. Consider the geometry or shape of diatomic molecules. Why do you think that the π_{2px} and π_{2py} orbitals have the same energy?

12. What are six insights your team gained about molecular orbitals by examining the two models?

Exercises

3. Place the electrons in the energy-level diagrams in **Model 2**, using B_2 and F_2 for Figures 3 and 4 respectively. Represent electrons with up and down arrows, and apply the Aufbau Principle, the Pauli Exclusion Principle, and Hund's Rule, as discussed in Activity 07-5 (Periodic Trends in Atomic Properties).

4. Using the insight you have gained from the energy level diagrams that you constructed in Exercise 3, identify which of the following statements are **correct** for B_2 and F_2.

a. Both molecules have unpaired electrons.

b. Only fluorine has unpaired electrons.

c. Both molecules have a bond order of 1.

d. The bond strength and bond energy are predicted, from the occupation of the orbitals, to be larger for F_2 than for B_2.

e. Both are homonuclear diatomic molecules.

f. Fluorine is paramagnetic. (See the **Information** section which directly follows the Problem on the next page.)

5. Some heteronuclear diatomic molecules have energy levels like those in Figure 3, while others have energy levels like those in Figure 4 (e.g., CN, CN^-, and CN^+).

a. Write the molecular orbital electron configurations for these three species. For example, the electron configuration for H_2 is $(\sigma_{1s})^2$.

b. Arrange these species in increasing bond order, increasing bond length, and increasing bond energy.

6. Identify the characteristic that determines whether a molecule is paramagnetic or diamagnetic.

7. Label the following species as *paramagnetic* or *diamagnetic*.

Li_2 _____ B_2 _____ Be_2 _____

CN _____ CN^- _____ CN^+ _____

PROBLEM

1. A convenient source of acetylene, C_2H_2, is a compound incorrectly called "calcium carbide," which suggests a formula of Ca_2C, composed of two Ca^{2+} ions and a C^{4-} ion. Calcium carbide actually has the formula CaC_2, and reacts with water to form calcium hydroxide and acetylene gas, which burns easily. Its actual formula suggests that it might contain two C^- ions, rather than one C^{4-}.

    Using a Lewis structure for acetylene, the correct formula for calcium carbide, and molecular orbital theory, predict the correct structure of calcium carbide. Propose a way to prove your prediction.

INFORMATION

Electrons behave like tiny bar magnets because they have a magnetic moment. If electrons are paired in molecular orbitals, the magnetic moments of the two electrons cancel each other out, because the magnetic fields point in opposite directions. Such a substance is *diamagnetic*. Unpaired electrons produce a net magnetic moment, and the molecule then is said to be *paramagnetic*.

GOT IT!

1. Identify which of the following statements are correct when molecular orbitals (MOs) are formed from two **1s** atomic orbitals, each centered on different H-atom nuclei (protons).

   a. Two bonding MOs are formed.

   b. Two antibonding MOs are formed.

   c. One bonding and one antibonding MO are formed, with the bonding MO having the lower energy.

   d. One bonding and one antibonding MO are formed, with the antibonding MO having the lower energy.

   e. If the MOs are occupied by electrons with the **same** value of m_s, both electrons will be in the bonding MO, as required by the Pauli Exclusion Principle and Hund's Rule.

Why?

The ideal gas law describes the relationship between the amount of a gas present and its pressure, volume, and temperature. These quantities are used in simple calculations, but they are also useful when dealing with chemical reactions involving gases or densities, or calculations that involve a known molar mass or volume of a gas. The ideal gas law is also used to predict or calculate changes in any of these attributes due to changes in the others. Any chemist who works with gases must understand and be able to use the ideal gas law.

Learning Objectives

- Understand the ideal gas law
- Master calculations involving the ideal gas law
- Learn how a molecular formula can be determined from the density of a gas

Success Criteria

- Be able to calculate each of the attributes of a gas (pressure, volume, amount, and temperature) when the others are given or known
- Correctly predict the results of changing one or more attributes of a gas
- Accurately calculate molar mass from gas density data and vice versa

Prerequisites

03-3   *Moles and Molar Mass*

03-4   *Determination of Chemical Formulas*

Information

The equation in the model below is called the *ideal gas law*. It gives an excellent approximation of the relationships between the attributes of a gas: pressure, P, in atmospheres; volume, V, in liters; amount, n, in moles, and temperature, T, in Kelvin. Studies of various gases have shown that the ratio PV/nT has a nearly constant value, for any gas. This value is called the *ideal gas constant* and is symbolized by the letter R. Gases for which $(PV)/(nT) = R$ are called *ideal gases*. Although this law neglects intermolecular forces of attraction and molecular volumes, it can usually be applied to all gases, as a reasonable approximation. The ideal gas law is commonly written in the form $PV = nRT$.

Since experiments show that, at one atmosphere of pressure and a temperature of 273.15 K (0 °C), one mole of an ideal gas occupies 22.414 L, the value of R is 0.08206 (L atm) / (mol K).

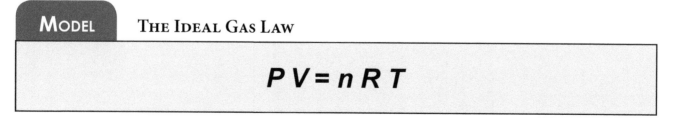

MODEL   THE IDEAL GAS LAW

$$PV = nRT$$

1. How would you explain the mathematical equation for the ideal gas law in words (rather than in symbols or abbreviations)?

2. According to the ideal gas law, if temperature increases while n and V remain unchanged, what must happen to the pressure? State your answer in words **and** provide an equation with four variables.

3. According to the ideal gas law, if the volume of a gas increases while n and T remain unchanged, what must happen to the pressure? State your answer in words **and** provide an equation with four variables.

4. According to the ideal gas law, if the number of molecules in a container increases while T and P remain unchanged, what must happen to the volume of the container? State your answer in words **and** provide an equation with four variables.

5. According to the ideal gas law, if the number of molecules in a container increases while T and V remain unchanged, what must happen to the pressure?

6. At what temperature does an ideal gas exert no pressure?

7. Describe a situation illustrating that it would be nonsense to use a Celsius temperature (such as 0 °C) for T in the ideal gas law.

EXERCISES

1. Calculate a value for R and determine the units of the ideal gas constant by using the following data for nitrogen gas:

$$n = 1.000 \text{ mole}, P = 482 \text{ atm}, V = 0.0121 \text{ L}, T = 295.15 \text{ K}$$

2. Calculate the volume occupied by 1.00 mole of a gas at 1.00 atm and 273.15 K. Remember this number, and note that the combination of these values for temperature and pressure is called *STP* (standard temperature and pressure).

GOT IT!

1. For the following situations, determine how quantities in the ideal gas law change to keep the ratio $(PV) / (nT)$ constant.

   a. If a gas in a sealed jar (V constant) is heated, and the temperature in Kelvin doubles, how large is the change in pressure?

   b. If a gas in a jar with a leaky seal (P and V constant) is heated, and its temperature (in Kelvin) quadruples, what fraction of the gas escapes?

   c. If the vapor in a cylinder in your automobile engine (T constant) is compressed by a piston to one tenth (1/10) of its previous volume, how large is the change in pressure?

2. According to the ideal gas law, which of the following is/are **correct** for a gas cylinder of fixed volume that is filled with one mole of oxygen gas?

   a. When the temperature of the cylinder increases from 15 °C to 30 °C, the pressure inside the cylinder doubles.

   b. When a second mole of oxygen is added to the cylinder, the ratio T/P remains constant.

   c. An identical cylinder filled with hydrogen gas under the same pressure contains more molecules, because hydrogen molecules are smaller than oxygen molecules.

   d. When a second mole of oxygen is added to the cylinder, the ratio T/P decreases by 50%.

   e. None of the above are correct.

PROBLEMS

1. An automobile tire is filled with air at a pressure of 44 lb/in² at 20 °C. The temperature drops to –5 °C. What is the new tire pressure in lb/in², assuming the volume of the tire does not change? Some students calculate a pressure of 0—is this correct? If not, what mistakes might they have made to obtain a pressure of 0 in their calculation?

2. The valve is opened between a 10 L tank containing a gas under 7 atm of pressure and a 15 L tank containing a gas at 10 atm. What is the final pressure in both tanks?

3. Derive an expression for the density of a gas (m/V), using the ideal gas law and the relationship between moles of a substance (n) and its mass (m). (M = molar mass).

4. A gaseous sample of a compound has a gas density of 0.647 g/L at 694 torr and 20 °C. What is the molar mass of this compound? If this compound contains only nitrogen and hydrogen, and is 82.2% nitrogen by mass, what is its molecular formula?

5. What is the density in g/L of neon gas at STP?

WHY?

The total pressure of a mixture of gases is the sum of the pressures that each component would exert if it were alone in the container. The ideal gas law relates the partial pressure caused by one component of a mixture to the amount of that component present in the container. Using partial pressures allows one to mix different amounts of different gases together, in order to tailor a mixture to specific requirements. Partial pressure is also useful for dealing with liquids that are in equilibrium with their own gaseous phase; e.g., air that is saturated with water at a given temperature.

LEARNING OBJECTIVES

- Understand the concept of partial pressure
- Master the use of partial pressure and mole fractions in calculations

SUCCESS CRITERION

- Perform accurate calculations involving partial pressures and mole fractions

PREREQUISITE

10-1 *The Ideal Gas Law*

INFORMATION

The pressure exerted by a gas in a container is caused by molecules of the gas hitting the walls of the container. If the different gases in a mixture do not interact with each other, as is the case for an ideal gas, then the collisions between one gas and the walls will not be affected by the presence of the other gases. The pressure due to each component of the mixture then will be independent of the pressures due to other components. The total pressure, P_T, is therefore the sum of the pressures of the different gases in the mixture (P_i).

$$\text{(Equation 1)} \quad P_T = \sum_i P_i \text{ where } P_i = n_i \frac{RT}{V}$$

In Equation 1, P_i is written in terms of the moles of that specific gas present, n_i, using the ideal gas law. The pressure of each component is called the *partial pressure* attributable to that component.

$$\text{(Equation 2)} \quad P_i = X_i P_T$$

Equation 2 shows that the partial pressure of the component i can also be calculated by multiplying the mole fraction of the given gas (indicated by the Greek letter chi, or X) by the total pressure. A *mole fraction* is defined as n_i/n_T, where n_i is the number of moles of the component gas i, and n_T is the total number of moles of gas in the mixture. X obviously ranges in value from 0 to 1. The sum of the mole fractions of all gases in the mixture is equal to 1.

GASES EXERT PRESSURE INDEPENDENTLY

Condition	Pressure	Volume	Temperature	Comment
initial	1.0 atm	22.4 L	273 K	1.0 mole of N_2 gas fills the tank.
final	3.0 atm	22.4 L	273 K	2.0 moles of O_2 gas have been added to the 1.0 mole of N_2 gas in the tank.

KEY QUESTIONS

1. What was the initial pressure caused by the nitrogen in the tank?

2. How much pressure do you think the nitrogen contributed to the final total pressure? Explain your reasoning.

3. If the sum of the pressure due to nitrogen and the pressure due to oxygen must equal the total pressure, what is the oxygen pressure in the model? Explain your reasoning.

4. What fraction of the total final pressure is due to nitrogen?

    Note: fraction due to nitrogen = N_2 pressure/total pressure = P_{N_2} / P_T

5. What is the mole fraction of nitrogen in the tank?

    Note: mole fraction of nitrogen = moles of N_2 / total moles = n_{N_2} / n_T

6. What is the relationship between the nitrogen pressure ratio in Key Question 4 and the nitrogen mole fraction in Key Question 5? Express this relationship as an equation.

INFORMATION

Your response to Key Question 6 is one way of stating *Dalton's Law of Partial Pressures.*

$$\frac{P_i}{P_T} = \frac{n_i}{n_T}$$

EXERCISES

1. The mole fraction of oxygen in air is 0.209. Assuming nitrogen is the only other constituent gas, what are the partial pressures of oxygen and nitrogen when the barometer reads 730 torr?

2. Calculate the partial pressure of 2.25 moles of oxygen that have been mixed (no reaction) with 1.75 moles of nitrogen in a 15.0 L container at 298 K.

PROBLEMS

(These problems require that you use what you have learned about the ideal gas law and stoichiometry in addition to Dalton's Law of Partial Pressures. The process of solving problems is what helps you to make connections between concepts and areas of understanding.)

1. A 30.0 L gas cylinder at 38 °C is filled with 3.33 moles of carbon dioxide and 3.66 moles of nitrogen. Calculate the total pressure, as well as the mole fractions and partial pressures of carbon dioxide and nitrogen, in the cylinder.

2. A balloon contains 0.3 moles of oxygen and 0.6 moles of nitrogen. If the balloon is at STP, what is the partial pressure of nitrogen?

3. Oxygen gas generated by the decomposition of potassium chlorate is collected over water at 25 °C, in a 1.00 L flask, and at a total pressure of 755.3 torr. The vapor pressure of water at 25 °C is 23.8 torr. How many moles of $KClO_3$ were consumed in the reaction?

$$2\ KClO_3(s) \rightarrow 2KCl(s) + 3O_2(g)$$

Why?

The kinetic molecular theory of gases is a simple model that serves to explain properties of gases, such as the relationships between pressure, volume, temperature, and amount of a gas, in terms of the properties of individual gas molecules. Through this theory, you will come to understand how some macroscopic properties of matter are a consequence of the microscopic properties of molecules. You will also be able to explain phenomena like diffusion and effusion, and appreciate why and how real gases differ from ideal gases. Finally, in examining the distribution of individual speeds of gas particles at a given temperature, you will understand temperature-dependent rates of reactions and reaction equilibrium positions. (In this discussion of gases, the expression "molecules in a gas" also refers to atoms in atomic gases, such as He.)

Learning Objective

- Understand and apply the postulates of the kinetic molecular theory

Success Criteria

- Correctly analyze properties of gases using kinetic molecular theory
- Perform accurate calculations involving average kinetic energy, root mean square velocity, and diffusion rates
- Compare distributions of molecular speeds at a given temperature

Prerequisite

10-1   *The Ideal Gas Law*

Model 1   Postulates of the Kinetic Molecular Theory

1. The volume of the individual molecules of a gas is negligible because their size is significantly smaller than the distance between them.

2. Gas molecules move constantly and randomly through the volume of the container they occupy, at various speeds and in every direction.

3. The forces of attraction and repulsion between molecules in a gas are negligible except when they collide with each other.

4. Gas molecules continually collide with each other and with the walls of the container they are in. The collisions are *elastic*, which means that the kinetic energy of the colliding molecules does not change.

5. The average kinetic energy of a collection of gas molecules, $<KE>$, is proportional to the absolute temperature of the gas. Specifically, using M = molar mass, $<v^2>$ = mean square molecular speed, R = gas constant, and T = Kelvin temperature:

$$<KE> = \frac{1}{2}M<v^2> = \frac{3}{2}RT$$

KEY QUESTIONS

1. In the kinetic molecular theory, what assumptions are made regarding the size of the molecules in a gas?

2. What assumption made regarding the interactions between molecules in a gas?

3. According to the kinetic molecular theory, are collisions between molecules visualized more like collisions between sponge balls, or collisions between bowling balls? Explain.

4. According to kinetic molecular theory, what information would you need in order to calculate the mean square molecular speed of a pure substance in the gas phase?

5. How can the postulates of the kinetic molecular theory be used to explain the ideal gas law; i.e., that the pressure of a gas is inversely proportional to the volume of the container it fills (V) and is directly proportional to the number of molecules (n) and the product of temperature and the gas constant (RT)?

EXERCISES

1. Illustrate the idea that gas molecules move constantly and randomly at various speeds and in every direction.

2. Calculate the average kinetic energy of 1 mol of hydrogen molecules and 1 mol of sulfur hexafluoride molecules at 25 °C. Calculate the average kinetic energy of 1 mol of hydrogen molecules and 1 mol of sulfur hexafluoride molecules at 30 °C.

3. Calculate the root mean square speed, $\langle v^2 \rangle^{1/2}$ of an O_2 molecule and a CH_4 molecule at 25 °C.

4. Compare the ratio of the root mean square speeds of O_2 and CH_4 at 25 °C to the ratio of their molar masses. Are these ratios equal or unequal? Explain.

5. You have a 5.0 L container filled with H_2 gas at standard temperature and pressure.

   a. Use the postulates of the kinetic-molecular theory to predict what happens when the conditions of the gas change as described in the table below. Fill in the boxes of the table with *increases*, *decreases*, or *stays the same*. Be sure you can justify your answers!

Condition	Average Kinetic Energy	Mean Square Velocity	Frequency of Collisions	Force per Collision
Temperature increases				
Volume increases				
Gas added				
H_2 replaced with N_2				

   b. Use your responses in the last row to explain why the pressure of a gas does not depend on its molar mass.

| MODEL 2 | DISTRIBUTIONS OF MOLECULAR SPEEDS OF GASES |

While the **average** kinetic energy of a gas sample is dependent on its temperature and molar mass, the molecular speeds or velocities of individual particles in a gas sample cover a wide range of values represented by a *Maxwell-Boltzmann distribution*. This is shown for a gas sample at three different temperatures in Figure 1.

Figure 2 illustrates the Maxwell-Boltzmann distributions of different gases at a single temperature.

Figure 1

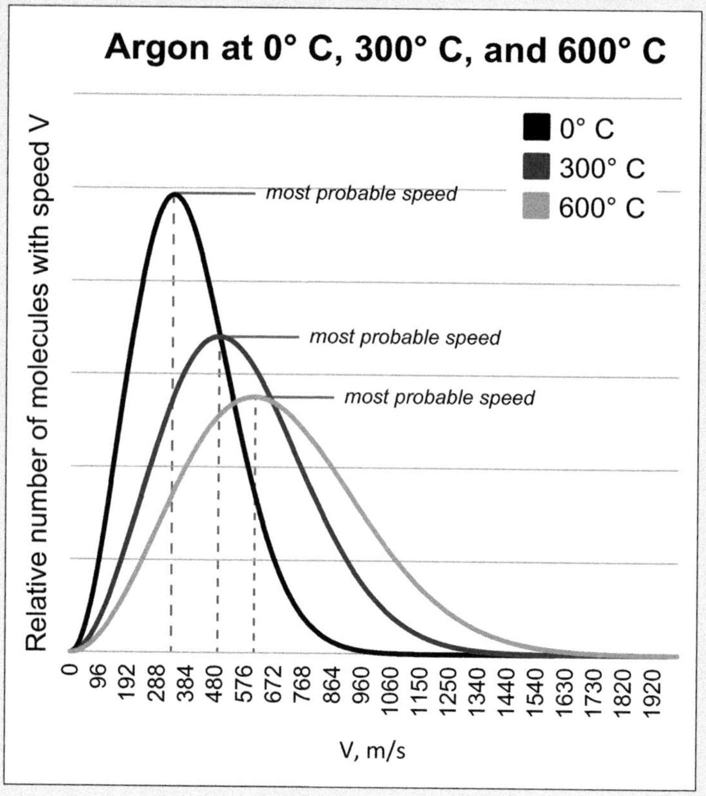

Argon at 0° C, 300° C, and 600° C

Figure 2

Neon, Argon, and Xenon at 300 C

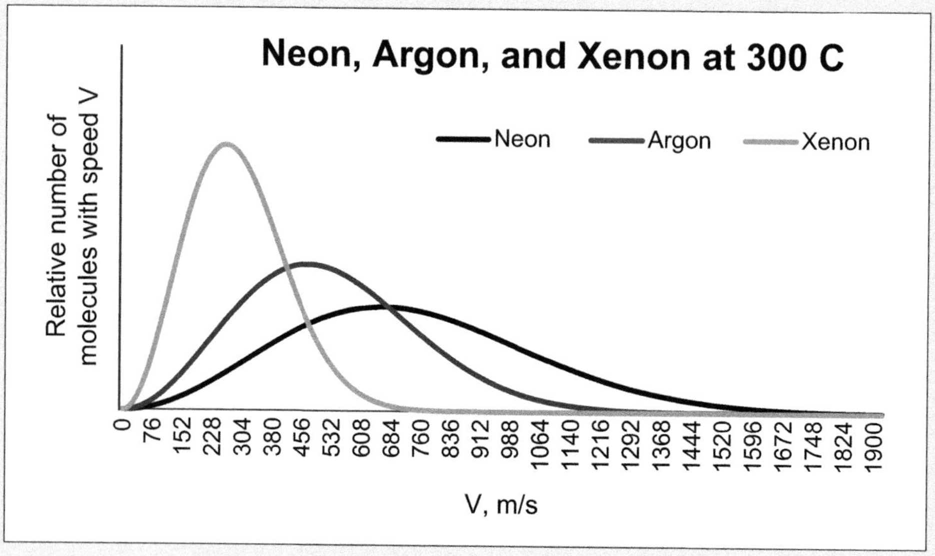

KEY QUESTIONS

6. What is the meaning of the height of a single distribution curve at a given point?

7. Assuming root-mean square velocity is close to the most probable velocity illustrated in Figure 1, how are the pictured distributions consistent with your previous calculations of molecular velocity?

PROBLEMS

1. Compare the root mean square velocity of a Kr atom at 25 °C with the speed of a jet airliner. Estimate the speed of a jet airliner based on your knowledge and experience.

2. *Diffusion* refers to the spread of one substance through another. The rate of diffusion depends on the speed at which the molecules of the substance are moving. Based on the kinetic molecular theory, does the rate of diffusion

   a.   increase or decrease as temperature increases? Explain.

   b.   increase or decrease as the mass of the molecules decreases? Explain.

3. Does the rate of diffusion change in direct proportion to changes in temperature or mass? Explain.

4. The rate of diffusion of a particular gas was measured and found to be 50.5 mm/min. Under the same conditions, the rate of diffusion of SF_6 was 18.7 mm/min. What is the molar mass of the unknown gas? What gas might this be?

5. One group of students in an introductory chemistry course is arguing that the pressures of a container of hydrogen gas and a container of sulfur dioxide gas, under the following conditions, cannot be equal. Both containers are sealed, with equal volume, at the same temperature. The students are arguing that the pressures cannot be equal because a sulfur dioxide molecule has a much larger mass than diatomic hydrogen, and therefore exerts a much greater force when it hits the walls of the container. They therefore conclude that the pressure of sulfur dioxide in the container should be larger than that of hydrogen. Their intuition seems to be really good. They simply ask, *"Would you rather be hit by a ping-pong ball or a baseball?"* They claim these are real and not ideal gases, so the ideal gas law is not relevant. Please write a note to them below, congratulating them if they are right, or, if they're not, explaining where they have gone wrong in their logic.

WHY?

Strong interactions between atoms, called *covalent bonding*, cause molecules to form. Electrostatic attractive forces between ions cause ionic bonds to form. Weaker interactions between atoms and molecules, called *noncovalent interactions*, cause gases to condense and form liquids and solids. These noncovalent interactions are also important in reactions involving enzymes, in the function of nucleic acids, and in determining the structure of proteins. The properties of liquids, solids, and solutions depend upon the nature and strength of the noncovalent interactions between their component molecules. With an understanding of the types of interactions between atoms, ions, and molecules, and their relative strengths, you can predict the physical properties of many materials and solutions.

LEARNING OBJECTIVES

- Develop an understanding of covalent and noncovalent interactions between atoms and molecules based on their relative magnitudes
- Understand the effects of noncovalent interactions on molecular structure
- Relate properties of materials to noncovalent interactions

SUCCESS CRITERIA

- Identify types of noncovalent interactions from molecular structure
- Predict properties of materials from molecular structure and concomitant noncovalent interactions

PREREQUISITES

- **03-1** *Nomenclature: Naming Compounds*
- **03-2** *Representing Molecular Structures*
- **08-3** *Valence Shell Electron Pair Repulsion Model*
- **08-4** *Electronegativity and Bond Polarity*

INFORMATION

Atoms and molecules interact with each other due to the attraction between opposite charges and the repulsion between like charges. Attractive interactions lead to the formation of molecules, liquids, and solids. When a solid melts or a liquid boils upon being heated, molecules acquire sufficient energy to overcome the attractive noncovalent forces and move away from each other. There is not enough energy, however, to overcome the covalent forces and break the bonds that hold the molecule together.

Interactions between atoms and molecules are given different names depending upon the characteristics or structural features of the atoms and molecules involved. The names are *covalent bonding, ionic, London dispersion, hydrogen bonding,* and *dipole-dipole*.

Covalent bonding interactions between atoms hold molecules together. These bonds are caused by atoms sharing electrons. Covalent bonds are strong, and it requires considerable energy to separate the atoms involved.

Ionic interactions result when electrons are not shared by atoms, but are transferred from one atom to another. The interaction is between a positively charged cation and a negatively charged anion.

London dispersion interactions occur in all substances. They result from fluctuations in charge distributions that cause one part of an atom or molecule to be momentarily positively charged and another part to be momentarily negatively charged. These instantaneous fluctuations result in attractive interactions. The noncovalent interactions between atoms and/or nonpolar molecules result only from the London dispersion interactions caused by these fluctuations.

In *hydrogen bonding*, a hydrogen atom bonded to one of the small electronegative atoms (oxygen, nitrogen, or fluorine) has a slight positive charge and is attracted to electrons on another of these electronegative atoms.

Dipole-dipole interactions are the interactions between permanent dipoles. Dipoles are formed in molecules because different atoms have different electronegativities. As a result, the electronegative end of the molecule attracts electrons and has a slight negative charge, and the opposite end has a slight positive charge. The dipoles in different molecules attract each other when the positive region of one molecule is close to the negative region of another. The dipoles repel each other when like charges are close together. Hydrogen bonding is a type of dipole-dipole interaction that is especially strong and localized to the specific atoms involved.

MODEL 1   INTERACTIONS BETWEEN ATOMS, IONS, AND MOLECULES

This table correlates the types of electrostatic properties of atoms, ions, and molecules with the attractive noncovalent interactions between oppositely charged particles that result from these properties. The strength of the attractive forces is greatest in the top left-hand corner and weakest in the bottom right-hand corner of the table. Examples are provided for each type of interaction.

Electrostatic Characteristics	Cation	H atom in Hydrogen Bond	Partial Positive Charge in Polar Molecule	Induced Partial Positive Charge in Nonpolar Molecule
anion	ionic bond	ion-dipole	ion-dipole	ion-induced dipole
	NaCl	H in H_2O–Cl^-	I in IF–Cl^-	Cl^-–Br_2
atom in hydrogen bond F, O, or N	ion-dipole	hydrogen bond	dipole-dipole	dipole-induced dipole
	Na^+–O in H_2O	H in H_2O–O in H_2O	O in H_2O–I in IF	O in H_2O–Br_2
partial negative charge in polar molecule	ion-dipole	dipole-dipole	dipole-dipole	dipole-induced dipole
	Na^+–F in IF	H in H_2O–F in IF	I in IF–F in IF	F in IF–Br_2
induced partial negative charge in nonpolar molecules	ion-induced dipole	dipole-induced dipole	dipole-induced dipole	London dispersion forces
	Na^+–Br_2	H in H_2O–Br_2	I in IF–Br_2	Br_2–Br_2

KEY QUESTIONS

1. The table includes interactions between atoms, ions, and molecules of pure substances and mixtures. Where do you find the interactions for pure substances on the table?

2. What is the strongest noncovalent attraction? What is the weakest?

3. What information is required to evaluate the strength of noncovalent interactions?

4. What is required for hydrogen bonding to occur between molecules?

Model 2 Identifying Interactions between Atoms and Molecules

One measure of the strength of the interaction between two atoms or molecules is the energy required to separate them. This energy is called the *dissociation energy*. Dissociation energies for some interacting systems are given below.

Figure 1

System		Dissociation Energy kJ/mol
2 Ne	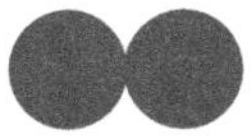	0.29
2 Xe	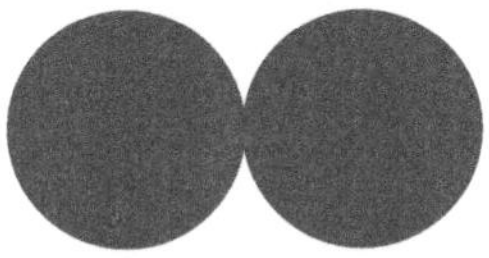	1.84
2 H_2O antiparallel dipoles	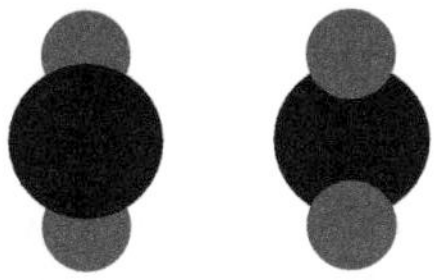	8.20
2 H_2O parallel dipoles	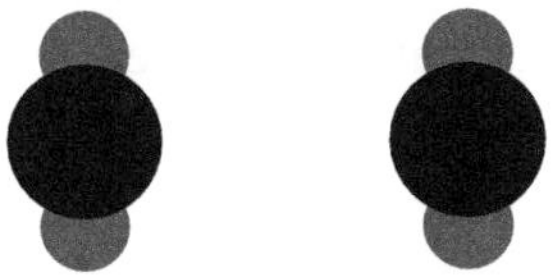	repulsive
2 H_2O head-to-tail	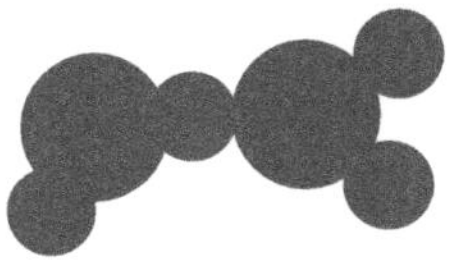	21.9
HO-H		427
NaCl	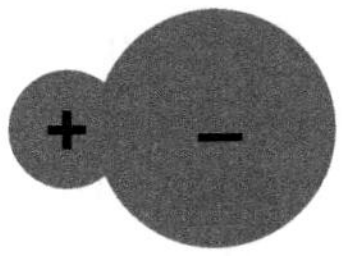	507

Key Questions

5. Which system in the model involves a repulsive interaction? Which system has the weakest interaction, according to its dissociation energy?

6. Of the systems in **Model 2**, which two have the strongest interactions?

7. As the value of the dissociation energy increases, more kinetic energy (reflected in temperature) is required to overcome the attractive forces between the particles. Which system in the model would you expect to have the lowest boiling point, and which would you expect to have the highest? Explain.

8. Why do you think more energy is required to separate two xenon atoms than is required to separate two neon atoms? Remember that electrical forces, which depend on the number and charges of the particles involved, are stronger than gravitational forces, which depend on the mass of the particles.

9. Why do you think that energy is required to separate two water molecules in an anti-parallel configuration, but in a parallel configuration, water molecules repel each other? (Remember that oxygen is more electronegative than hydrogen. This means that the oxygen in water is slightly negatively charged, and the hydrogen is slightly positively charged.)

10. Why do you think that water in a head-to-tail orientation has the highest dissociation energy, as compared to the anti-parallel and parallel orientations?

11. What insight has your team obtained so far about why some substances are gases at room temperature, while others are liquids or solids?

GOT IT!

1. Oxygen, O_2, and methanol, CH_3OH, have the same mass: 32 amu. Explain why oxygen is a gas at room temperature, while methanol is a liquid at room temperature.

EXERCISES

1. Compare the systems in the table below with those in the Model and identify the dominant interaction as one of the following: *London dispersion, covalent bonding, hydrogen bonding, dipole-dipole,* or *ionic.*

System	2 Ne	2 Xe	2 H₂O anti-parallel
Interaction Type			

System	2 H₂O parallel	2 H₂O head-to-tail	HO-H
Interaction Type			

2. The boiling points of butane, acetone, and propanol are 0 °C, 56 °C, and 97 °C, respectively. These molecules, shown below, have about the same number of atoms and electrons, and similar molar masses,, yet their boiling points are quite different. Identify the dominant noncovalent interaction present for each of these molecules and explain why these interactions cause their boiling points to be so different.

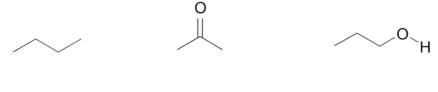

3. Hexane, 2-pentanone, and 1-pentanol are larger than the molecules mentioned in Exercise 2. Their boiling points are 69 °C, 102 °C, and 137 °C, respectively. For each of the types of interactions that you identified in Exercise 2, explain why the boiling point increases with the size of the molecule. Use your observations of the way neon and xenon atoms interact, along with the fact that xenon has more electrons than neon, to help you identify which interaction increases with the number of electrons in the molecule.

4. Identify the most important type of noncovalent interaction for each of the following substances: $BaSO_4$ (a salt), Ar, NO_2, and HF.

5. Explain why the following compounds have different boiling points:

HF (20 °C)   HI(−35 °C) $TiCl_4$ (nonpolar, 136 °C)   NaCl (1,413 °C)

6. Identify which of the following does not form hydrogen bonds: HCl, H_2O, C_2H_6, and NH_3. Explain.

Got It!

2. Do you expect that adding an electronegative atom to an alkane will increase or decrease its boiling point? Explain.

3. Why is the boiling point of ammonia higher than the boiling point of ethane?

4. What type of interaction is responsible for sodium chloride's very high melting point?

5. What type of interaction is responsible for water's high boiling point?

Phase Changes in Pure Substances

Why?

A phase diagram is a convenient way to represent which phase (solid, liquid, or gas) or combination of phases is present at different temperatures and pressures. Knowing the relationship of temperature and pressure to a material's phase is necessary in order to determine its suitability for a particular application (e.g., a high-temperature engine component or a cooking oil). Phase diagrams are used, for example, to plan the synthesis of artificial diamonds and new materials, or to deduce the history of geological samples. Additionally, knowledge of a material's capacity to absorb energy in a particular phase or while changing phases is essential to the temperature and stability of systems involving chemical reactions and energy transfer.

Learning Objectives

- Interpret the major features of a phase diagram
- Interpret a heating or cooling curve to determine the physical and thermochemical properties of a pure substance

Success Criteria

- Identify stable phases, at particular temperatures and pressures, from a phase diagram
- Use a phase diagram to identify the temperatures and pressures at which phase changes occur
- Use a heating curve to determine a substance's melting and boiling points, specific heat capacities for each phase, and enthalpies associated with phase changes

Prerequisites

06-1 *Thermochemistry and Calorimetry*

06-2 *Internal Energy and Enthalpy*

Information

A *phase diagram* summarizes the relationships between a substance's phase, temperature, and pressure. It represents a map of the states of matter for a pure substance. Within a pure phase, the addition or removal of heat results in a change in temperature.

Boundary lines between pure phases represent the temperatures and pressures at which two phases are in equilibrium. As a phase change or transition occurs across these boundary lines, the temperature does not change until the phase change is complete.

A *triple point* marks the temperature and pressure where three boundary lines come together. At these points, three phases are in equilibrium.

A phase diagram can be produced by observing the which phase of a substance is present under specific temperature and pressure conditions, as heat energy is added to or removed from the system. A *heating curve* is a plot of the temperature of the substance as a function of the energy added. A *cooling curve* is a plot of the temperature as a function of the energy removed.

Measurement of the amount of energy transferred in a heating or cooling experiment makes it possible to determine the enthalpies of fusion and vaporization for phase changes, as well as the specific heat capacities of different pure phases.

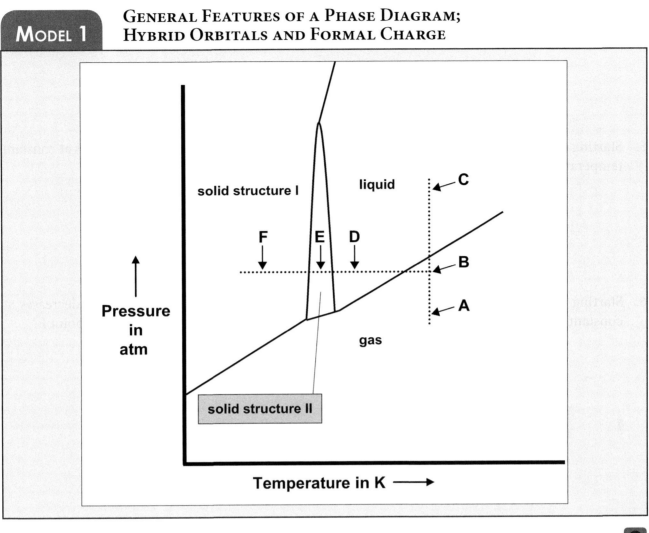

KEY QUESTIONS

1. What quantities are plotted on the *x*- and *y*-axes of a phase diagram?

2. a. How many different phases are shown by the phase diagram in the model?

   b. What are the labels used to identify them?

3. What do the solid lines in a phase diagram represent?

4. a. How many triple points are visible in the phase diagram in **Model 1**?

b. What three phases are in equilibrium at each of them?

5. Starting at point A in the model phase diagram, what happens as the pressure increases at constant temperature, following the dotted line to point C?

6. Starting at point B in the model phase diagram, what happens as the temperature decreases at constant pressure, following the dotted line to point D, then to point E, and finally to point F?

MODEL 2	GENERAL FEATURES OF A HEATING CURVE

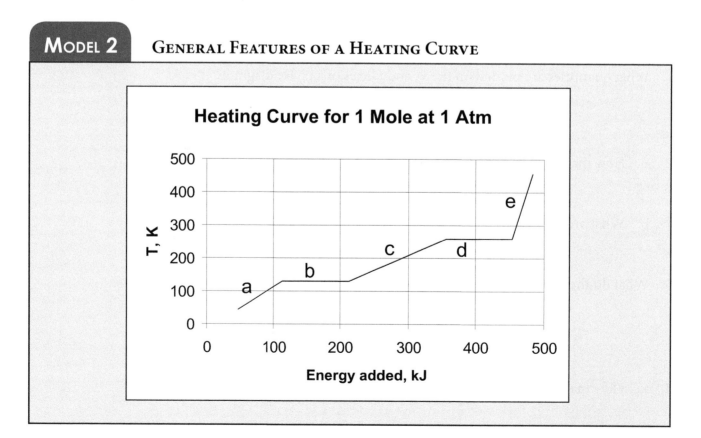

7. What quantities are plotted on the *x*- and *y*-axes in a heating curve?

8. Which of the five line segments (labeled a through e) in the heating curve (**Model 2**) correspond to the processes below?

_____ heating the solid   _____ heating the liquid   _____ heating the gas

phase transition
_____ between solid and liquid

phase transition
_____ between liquid and gas

EXERCISES

1. The *critical point* in a phase diagram is where the boundary line between the liquid and gas phase ends. Find and label the critical point in the phase diagram in **Model 1**.

2. Sketch a phase diagram for methanol (CH_3OH) using the data in the following table.

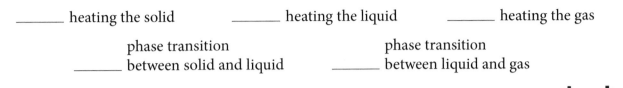

Item	Temperature	Pressure
solid-liquid-gas triple point	−97.7 °C	1.07×10^{-6} atm
melting point	−97.6 °C	1.0 atm
boiling point	64.7 °C	1.0 atm
critical point	240 °C	78.5 atm

3. Use the heating curve in **Model 2** to estimate both the melting and boiling points of the substance.

4. Use the heating curve in **Model 2** to estimate values for the following parameters of the substance. Assume that the molar mass of the substance is 120 g/mol.

   a. molar heat of fusion

   b. molar heat of vaporization

   c. specific heat capacity of the solid

   d. specific heat capacity of the liquid

   e. specific heat capacity of the gas

Why?

Molecules in the liquid phase escape to the gas or vapor phase if they have sufficient kinetic energy. At the same time, vapor phase molecules return to the liquid phase if they don't have sufficient kinetic energy and are trapped in the liquid phase by intermolecular forces of attraction. This is an example of a phase equilibrium process. The distribution of molecular kinetic energies within a system is dependent upon the temperature of the system. Further, the **extent** of equilibrium within the system is **quantitatively dependent** on the temperature of the system. A familiar example of this type of equilibrium is the vapor pressure of water. Everyday experience tells us that the air can contain much more water vapor at high temperatures than at low temperatures. The condensation of water as dew as the temperature drops during the night is a result of this temperature-dependent phase equilibrium process.

Learning Objectives

- Recognize the factors involved in liquid phase – vapor phase equilibrium processes
- Understand the quantitative relationships between vapor pressures, temperatures, and enthalpy of vaporization

Success Criterion

- Ability to use the Clausius-Clapeyron equation to relate vapor pressures, temperatures, and enthalpy of vaporization for a given liquid

Prerequisites

06-3 *Hess's Law: Enthalpy is a State Function*

10-2 *Partial Pressure*

10-3 *Kinetic Molecular Theory of Gases*

11-1 *Interactions between Atoms and Molecules*

11-2 *Phase Changes in Pure Substances*

Model 1 Liquid – Vapor Phase as an Equilibrium Process

The normal boiling point of a liquid is the temperature at which the vapor pressure, P, is equal to an ambient total pressure of 1 atmosphere (atm) (which is defined as 760 mm Hg or 760 torr). At temperatures lower than the normal boiling point, the vapor pressure, P, is lower than 760 torr (but not zero, due to the equilibrium established between the liquid phase and the vapor phase). Conversely, if the total pressure is lowered by applying a vacuum, the boiling point of the liquid decreases. For water, this equilibrium is written as:

$$H_2O(l) \rightleftharpoons H_2O(g) \qquad \Delta H_{vap} = 40.68 \text{ kJ/mol}$$

The enthalpy of vaporization is, of course, endothermic, at 40.68 kJ/mol.

Figure 1 shows measured vapor pressures of several volatile liquids as a function of temperature, and can be considered as a small part of the overall phase diagram.

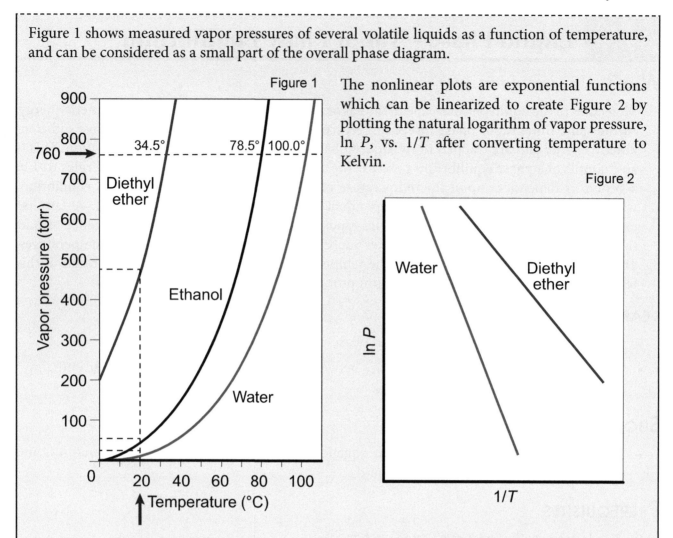

Figure 1

The nonlinear plots are exponential functions which can be linearized to create Figure 2 by plotting the natural logarithm of vapor pressure, ln P, vs. $1/T$ after converting temperature to Kelvin.

Figure 2

The second type of plot is the basis for the Clausius-Clapeyron equation (Equation 1):

$$\text{(Equation 1)} \quad \ln P = \left(\frac{-\Delta H_{vap}}{R}\right)\left(\frac{1}{T}\right) + C$$

where: R is the universal gas constant of 8.314 J/K•mol

ΔH_{vap} is in J/mol T is in Kelvin C is a constant

This form of Equation 1 is very similar to others you will encounter later in your studies of chemistry. The nature of the constant C will be explained more fully in discussions of changes in free energy and equilibrium constants.

Since this is a linear equation, it has the general form $y = mx + b$, so it can be shown that any two points of related temperature and pressure for a given liquid can be described quantitatively by Equation 2:

$$\text{(Equation 2)} \quad \ln\left(\frac{P_2}{P_1}\right) = \left(\frac{-\Delta H_{vap}}{R}\right)\left(\frac{1}{T_2} - \frac{1}{T_1}\right)$$

KEY QUESTIONS

1. What are the normal boiling points of diethyl ether, ethanol, and water? How do you know?

2. Which liquid has the highest vapor pressure at 20 °C?

3. In Equation 1, what is the slope of the linear function?

4. Based on the second plot and your response to Key Question 3, which liquid has the higher value of ΔH_{vap}? Explain.

EXERCISES

1. Use data from the Model to estimate the enthalpy of vaporization of diethyl ether.

2. What would be the boiling point, in °C, of water under partial vacuum conditions of 205 torr?

11-4 Cubic Unit Cells of Crystalline Solids

WHY?

A unit cell is the smallest group of atoms or molecules in a crystal. If many unit cells are stacked together in three dimensions, without any gaps, they produce their source crystal. The structures of crystalline solids are described in terms of unit cells, and you will find that structure is the key to understanding many properties of crystalline solids. There are many types of unit cells, each with specific geometric properties, and this introduction will focus on three of these, all related to the geometry of a perfect cube. Determination of the detailed structures of unit cells of crystalline materials is carried out using X-ray crystallography. X-ray crystallography is not only applicable to simple structures like metals and simple salts; it can also be used to examine much larger molecules like proteins and DNA helices, as long as they can be crystallized.

LEARNING OBJECTIVES

- Identify cubic unit cells by their names and characteristics
- Determine the relationships between atomic size, unit cell characteristics, and density of the material

SUCCESS CRITERIA

- Accurately determine the number of atoms in a particular cubic unit cell
- Learn to quantitatively relate the density of a crystal, the radius of the atoms involved, and its unit cell composition, structure, and size

INFORMATION

You might expect that any number of unit cells could produce crystal structures, but in 1848 the French physicist August Bravais proved that only fourteen distinct unit cells, now called *Bravais lattices*, are needed to generate all the possible three-dimensional crystal lattices. In this activity, three of these fourteen will be considered: primitive cubic, body-centered cubic, and face-centered cubic. A cubic unit cell has sides of equal lengths that intersect at 90° angles. Many metals form solids with these cubic structures. When describing the structure of a metallic crystal, atoms are represented by hard spheres packed next to and on top of each other. The different ways these hard spheres, or balls, can be stacked in the unit cell are responsible for the variation between types.

MODEL UNIT CELL CUBICS

r = metallic radius of an atom   a = length of a side of the cubic unit cell

Primitive Cubic
The atoms are touching along an edge.

$a = 2r$

1/8 of an atom

Face-Centered Cubic (aka Cubic Close-Packed)
The atoms are touching along a face diagonal.

$a = \sqrt{8}r$

1/8 of an atom

1/2 of an atom

1/2 of an atom

Body-Centered Cubic
The atoms are touching along a body diagonal.

$a = \dfrac{4}{\sqrt{3}}r$

1/8 of an atom

whole atom

KEY QUESTIONS

1. What are the names of the three unit cells in the **Model**?

2. What is the relationship between the radius of a metallic atom and the length of a side of the cube it forms for each of the unit cells in the **Model**?

3. When an atom is located at the corner of a unit cell, why does the cell only contain 1/8 of it? When it is in the center of a side of the cell, why does the cell contain 1/2 of it?

4. How many atoms are associated with each of the unit cells in the **Model**? (Sum the fractional and complete atoms.)

5. For each of the three unit cells in the **Model**, what is the relationship between the radius of the atom and the volume of the unit cell?

6. How can you calculate the mass of the atoms in a unit cell, given the molar mass of the element that composes the crystal?

7. Keeping in mind your answers to Key Questions 5 and 6, how can you calculate the density of a metal, given the characteristics of the unit cell?

EXERCISES

1. How many atoms are there in a unit cell of lithium, which crystallizes in a body-centered cubic structure?

2. Use the Pythagorean theorem, when necessary, to show that the relationships given in the model between the metallic radius of an atom and the side of a unit cell are correct.

3. Determine the metallic radius of a copper atom from x-ray diffraction data which reveal that the structure is face-centered cubic with a unit cell length of 361.6 pm.

4. Using the x-ray diffraction data, given in Exercise 3, determine the density of a copper crystal.

PROBLEMS

1. Copper crystallizes in a face-centered cubic structure. What is the fraction of the resulting unit cell volume that is occupied by copper atoms, assuming the hard sphere model for the atoms with a radius of 127.8 pm? Does this volume depend upon the composition of the crystal, or is it the same for all face-centered cubic structures?

2. Nickel crystallizes in a face-centered cubic structure with a density of 8.908 g/cm^3. What is the metallic radius of nickel, given this information?

3. What is the density of platinum (which crystallizes in a face-centered cubic structure and has a metallic radius of 139 pm)?

4. Sketch the NaCl unit cell, using the following information: It has chloride ions in a face-centered cubic arrangement, with the centers of sodium ions positioned along each edge, between the chloride ions. There is an additional sodium ion in the center position that cannot be seen. This is known as the "rock salt" structure.

   What fraction of each edge-positioned sodium ion lies within the volume of the cube? How many of the edge-centered atoms are there? How many sodium ions total are found within the unit cell, including these edge-positioned ions and the one sodium ion in the center? Is this consistent with the formula of NaCl?

WHY?

A *solution* is a homogeneous combination of two or more substances. The major component is called the *solvent*; the other components are the *solutes*. Knowledge of how we measure concentration, the factors that determine solubility, and the unique properties of solutions enables scientists, engineers, medical professionals, and you, too, to predict properties of unfamiliar solutions.

LEARNING OBJECTIVES

- Identify factors that affect solubility
- Define measures of solution composition
- Learn how a solute affects the vapor pressure, boiling point, freezing point, and osmotic pressure of a solvent

SUCCESS CRITERIA

- Predict solubilities of various substances and situations
- Calculate solution concentrations using different units
- Correctly predict the changes a solute will cause to a solvent's properties

PREREQUISITES

04-3 *Solution Concentration and Dilution*

10-2 *Partial Pressure*

11-1 *Interactions between Atoms and Molecules*

TASKS

Use your textbook as a reference as you perform the following tasks:

1. Complete Table 1: Factors Affecting Solubility
2. Complete Table 2: Measures of Solution Concentration
3. Complete Table 3: Solute Effects

Table 1 Factors Affecting Solubility

Note: In Table 1, identify general trends. Exceptions do sometimes occur; we will learn about exceptions and the reasons behind them later.

Condition	Nonbonding Interaction (if applicable)	Effect on Solubility
polar solute/polar solvent	*Dipole-dipole*	*The solute dissolves.*
nonpolar solute/nonpolar solvent		
polar solute/nonpolar solvent		
nonpolar solute/polar solvent		
increasing the pressure of a gas over a liquid solvent		
increasing the temperature of a dissolving solid		
increasing the temperature of a dissolving gas		

Table 2 Measures of Solution Concentration

Quantity	Formula	Example
mass percent	$= \dfrac{\text{solute mass}}{\text{total soln mass}} \times 100\%$	20 g of NaCl in 90 g water = (20 g NaCl / 110 g soln) × 100% = 18% by mass
mole fraction		0.25 mol NaCl in 5.50 moles water = 0.25 mol NaCl / 5.75 total moles = 0.043 mole fraction

Quantity	Formula	Example
molarity (M)		0.25 mol NaCl in 0.50 L of solution = 0.25 mol NaCl / 0.5 liter soln = 0.50 M
molality (*m*)		0.25 moles NaCl in 1.5 kg of water = 0.25 mol NaCl / 1.5 kg water = 0.17 *m*

In Table 3, identify the qualitative effect that the addition of a solute has on the specified solvent property.

Table 3 Solute Effects	
Property	**Qualitative Effect of Added Solute**
vapor pressure	*reduces vapor pressure*
boiling point	
freezing point	
osmotic pressure	*increases osmotic pressure*

KEY QUESTIONS

1. What general rule regarding the solubility of polar/nonpolar solutes in polar/nonpolar solvents can you identify in Table 1?

2. What is the difference between a solution of sugar and water that is 1 M and one that is 1 *m*, according to Table 2?

3. Why do people put salt on ice in the winter?

4. Is there a correlation between the effect of a solute on the vapor pressure of a solvent and its effect on the solvent's boiling point? Explain.

EXERCISES

1. Which solvent, water (H_2O) or hexane (C_6H_{14}), would you use to dissolve the following solutes? Explain.

   a. $(NH_4)_2SO_4$

   b. HF

   c. octane (C_8H_{18})

2. A solution is prepared by dissolving 20.00 g of acetic acid (CH_3COOH) in 800.0 g of water. The density of the resulting solution is 1.095 g/mL.

    a. What is the mass percent of acetic acid in the solution?

    b. What is the molarity of the solution?

    c. What is the molality of the solution?

    d. What is the mole fraction of acetic acid in the solution?

12-2 Colligative Properties

WHY?

Properties of a liquid that depend upon the relative number of other molecules dissolved in it, but not the identity of these molecules, are called *colligative properties*. You should understand the quantitative as well as the qualitative nature of these properties, as they are important in everyday life as well as in scientific research. For example, salt is thrown on ice to melt it, salty water is used to boil eggs in Denver, and osmotic pressure causes the flow of water through plants, even to the very tops of trees.

LEARNING OBJECTIVES

- Identify the basic colligative properties
- Determine the effects of solutes on the boiling point, freezing point, and osmotic pressure of a solvent

SUCCESS CRITERIA

- Accurately calculate freezing point depression, boiling point elevation, and osmotic pressure
- Determine the molar mass of a solute based on its effect on the colligative properties of a solvent

PREREQUISITES

04-3   *Solution Concentration and Dilution*

12-1   *Solutions*

INFORMATION

Colligative properties of a solution include the lowering of vapor pressure, the depression of the freezing point, the elevation of the boiling point, and the osmotic pressure caused by adding a nonvolatile solute to a solvent. These properties are all based on the idea that the increased attractive forces between solvent and solute result in solutions that are more stable (compared to pure solvent), which leads to solvent molecules being retained in the liquid phase.

Since the volume of a solution changes with changes in temperature, the molarity of the solution will change as well. *Molality* is therefore the concentration expression used, because it involves mass instead of volume, and is consequently unaffected by temperature.

MODEL 1	NOTATION AND EQUATIONS FOR COLLIGATIVE PROPERTIES

ΔT_b = elevation of the boiling point

K_b = molal boiling point elevation constant

c_m = molality of the solution = moles solute/kg of solvent

ΔT_f = depression of the freezing point

K_f = molal freezing point depression constant

Π = osmotic pressure

R = the ideal gas constant = 0.08206 L atm/mol K

T = temperature in K

RT = a measure of the average translational energy of a collection of molecules

M = molarity of the solution

i = van 't Hoff factor, which is the ratio of the moles of particles in solution to the moles of solute dissolved

Units of pressure: 1 atmosphere = 760 torr

The boiling point is elevated: $\Delta T_b = i K_b c_m$

Freezing point is depressed: $\Delta T_f = i K_f c_m$

Solvent will move from the region of low solute molarity to the region of high solute molarity, even through a separating membrane, producing an osmotic pressure: $\Pi = i M R T$

KEY QUESTIONS

1. What colligative properties are included in this activity?

2. Why is molality, rather than molarity, used as the measure of concentration in the equations for freezing point depression and boiling point elevation?

3. How can you determine the molar mass of a polymer by measuring the osmotic pressure produced by dissolving some amount of said polymer in a solvent?

EXERCISES

1. Explain why the freezing point depression of the following compounds in water (each 1.00 m solutions) increases in the indicated order.

$$\text{sucrose} < HOBr < LiBr < CaCl_2$$

2. Determine the boiling point of 1 L of water (K_b = 0.51 °C/m) when 1 oz of salt (28 g sodium chloride) is added to it. Does it make sense for people in Denver to add salt to increase the boiling point of water in their mile-high city? Explain.

3. Determine the freezing point of 1 L of water (K_f = 1.86 °C/m) when 1 oz of salt (28 g sodium chloride) is added to it.

4. Arrange the following aqueous solutions in order of increasing osmotic pressure:

    0.15 M NaBr, 0.02 M glucose, 0.40 M Na_2SO_4, 0.50 M HF, and 0.02 M $MgCl_2$.

5. 3.8 g of polyethylene is dissolved in benzene to produce 150.0 mL of solution. The osmotic pressure is found to be 6.70 torr at 25 °C. Calculate polyethylene's average molar mass.

6. Sulfur exists in many forms, all with the general molecular formula S_n. If 0.48 g of sulfur are added to 200 g of carbon tetrachloride, and, as a result, the freezing point of the carbon tetrachloride ($K_f = 30\,°C/m$) is depressed by $0.28\,°C$, what is the molar mass and molecular formula of the sulfur?

7. Beaker A contains 100.0 mL 1.0 M salt solution (NaCl). Beaker B contains 100.0 mL pure water. Both beakers are placed inside a large container, which is sealed.

   a. Draw a graph showing how both volumes (solution in beaker A and liquid in beaker B) change with time. Explain.

b. Draw a graph showing how both concentrations (the solution in beaker A and the liquid in beaker B) change with time. Explain.

ACTIVITY
13-1　Rates of Chemical Reactions

WHY?

In chemical kinetics, we look at the rate at which reactants are converted into products. An *experimental rate law* can be determined from such measurements, and this rate law can be used to deduce the mechanism of a reaction (how it occurs on the molecular level). Knowledge of the reaction mechanisms and the factors that affect the rate of a reaction makes it possible for a chemist to plan the efficient and cost-effective production of industrial, pharmaceutical, and consumer chemicals, or to understand intricate biochemical processes. In order to understand reaction rates, it is essential to understand **how** reactions occur.

LEARNING OBJECTIVE

- Understand reaction rates and rate laws

SUCCESS CRITERIA

- Use kinetic data to identify a reaction as zero, first order, or second order
- Determine the rate constant for a reaction from kinetic data
- Graph the concentrations of reactants and products as a reaction progresses

PREREQUISITES

04-1　*Balanced Chemical Reaction Equations*

04-3　*Solution Concentration and Dilution*

12-1　*Solutions*

INFORMATION

The *average rate* of a reaction is given by the change in the concentration of a reactant, Δc_r, or a product, Δc_p over some time interval $\Delta t = t_{final} - t_{initial}$ (provided that the stoichiometric coefficient for the product and reactant is 1). If the stoichiometric coefficient for some species is not 1, then its rate of reaction is determined by dividing the concentration of that species by its stoichiometric coefficient.

$$\text{(Equation 1)} \quad \text{reaction rate} = -\frac{\Delta c_r}{\Delta t} = \frac{\Delta c_p}{\Delta t}$$

Since the reaction rate is always given as a positive number and the reactant concentration is decreasing, $\dfrac{\Delta c_r}{\Delta t}$ has been multiplied by –1 in Equation 1.

The *instantaneous rate* of a reaction at a particular time is the limit of the average rate at that time as Δt approaches 0, and is tangent to the curved plot of concentration vs. time at that point. If the average rate is calculated over a small time interval, it is a good approximation of the instantaneous rate.

The *initial rate* of a reaction is the instantaneous rate at $t = 0$, when the reactants are mixed together.

293

MODEL 1   KINETICS OF OZONE DECOMPOSITION

Under certain conditions, ozone in the atmosphere decomposes by dissociation:

$$(O_3 \rightarrow O_2 + O)$$

The concentration of ozone as a function of time is given by the graph and data table below. Note that the curved line in the graph refers to the left scale, which gives the ozone concentration. The straight line refers to the right scale, which gives the natural logarithm of the ozone concentration divided by the initial ozone concentration.

Figure 1

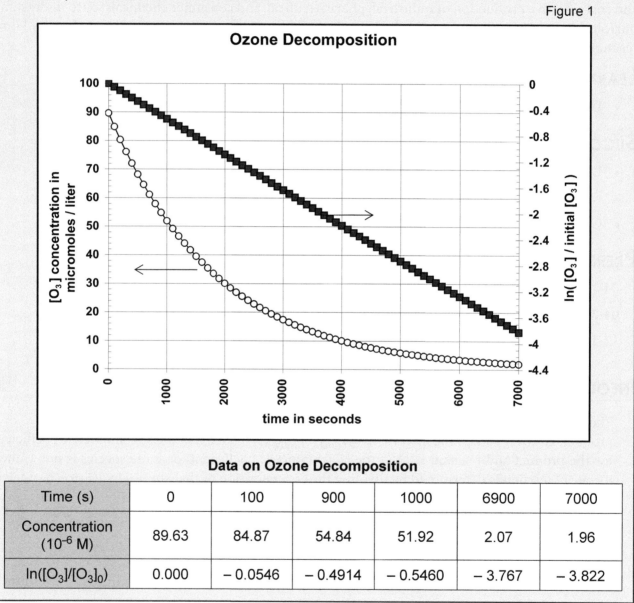

Data on Ozone Decomposition

Time (s)	0	100	900	1000	6900	7000
Concentration $(10^{-6}$ M)	89.63	84.87	54.84	51.92	2.07	1.96
$\ln([O_3]/[O_3]_0)$	0.000	-0.0546	-0.4914	-0.5460	-3.767	-3.822

KEY QUESTIONS

1. What quantities are plotted in the graph shown in **Model 1**?

2. What is the relationship between the data in the table and the data displayed in the graph?

3. In the model, how long does it take for half of the ozone to decompose? What is the concentration of ozone at that time? (This time, Δt, is called the half-lifetime, or *half-life*, of ozone.)

4. What is the reaction rate of the ozone decomposition reaction? Provide both the magnitude and the units.

5. What is the rate of the reaction after 6900 seconds have passed?

6. What happens to the reaction rate as the concentration of ozone decreases?

INFORMATION

A *rate law* for a reaction, sometimes called the *differential rate law*, indicates how the rate of the reaction depends on the concentrations of the chemical reactants (and sometimes products) involved at any given time during the reaction. Generally, rate laws take the following form: the rate of the reaction equals some constant multiplied by a product of concentrations $[A]$, $[B]$, and $[C]$, with exponents x, y, and z, respectively. In equation form:

$$\text{rate} = k\,[A]^x\,[B]^y\,[C]^z$$

An exponent gives the *order* of the reaction with respect to that chemical species. For example, if x and z equal 1, then the reaction is first order with respect to A and C, and if $y = 2$, it is second order with respect to B. The *overall order* of the reaction is the sum of the exponents in the rate law; for this example, $1 + 2 + 1 = 4$.

The order of a reaction often differs from its stoichiometric coefficients, which describe the numbers of reactant and product molecules involved in the reaction. The order of a reaction depends on the reaction **mechanism**, not on the number of molecules involved in the overall reaction.

MODEL 2 RATE LAWS FOR CHEMICAL REACTIONS

In the equations below, the reactant is represented by A. $[A]_t$ represents the concentration of A at time t, $[A]_0$ represents the initial concentration of A. k is the *rate constant*; and *ln* is the natural logarithm. The rate constant is the constant of proportionality in the rate law. The *integrated rate law* is obtained from the rate law by integration of the differential equation.

Zero order reaction

A → products

rate law $\qquad\qquad$ $\text{rate} = -\dfrac{\Delta[A]}{\Delta t} = k[A]_0$

integrated rate law $\quad$ $[A]_t = -kt + [A]_0$

First order reaction

A → products

rate law $\qquad\qquad$ $\text{rate} = -\dfrac{\Delta[A]}{\Delta t} = k[A]$

integrated rate law, exponential form $\qquad$ $[A]_t = [A]_0\, e^{-kt}$

integrated rate law, logarithmic form $\qquad$ $\ln\left([A]_t\right) = -kt + \ln[A]_0$

Or, equivalently, $\qquad$ $\ln\left(\dfrac{[A]_t}{[A]_0}\right) = -kt$

Second order reaction

A → products

rate law $\qquad\qquad$ $\text{rate} = -\dfrac{\Delta[A]}{\Delta t} = k[A]^2$

integrated rate law $\qquad$ $\dfrac{1}{[A]_t} = kt + \dfrac{1}{[A]_0}$

KEY QUESTIONS

7. What are four similarities between zero, first, and second order reactions, as represented by **Model 2**?

8. What are three differences between zero, first, and second order reactions, as represented by **Model 2**?

9. If the rate constant k increases when temperature increases, what happens to the rate of the reaction, according to the rate laws in **Model 2**? Is your answer the same for zero, first, and second order reactions?

10. If the concentration of reactant A increases, what happens to the rate of the reaction, according to the rate laws in **Model 2**? Is your answer the same for zero, first, and second order reactions?

11. If time is on the x-axis, what do you need to plot on the y-axis in order to obtain a straight line for each of the reaction orders in **Model 2**? To answer this question, compare the integrated rate laws to the general equation for a straight line, $y = mx + b$, where m is the slope and b is the y-intercept.

    a. a zero order reaction.

    b. a first order reaction.

    c. a second order reaction.

12. What are the slope and y-intercept of the straight line that you identified in your answer to Key Question 11 for:

    a. a zero order reaction?

    b. a first order reaction?

    c. a second order reaction?

EXERCISES

1. On the graph in **Model 1**, draw a vertical line to mark the half-life of ozone, and label this line $t_{1/2}$. Draw a horizontal line on the graph in **Model 1** at half the initial concentration, and label this line $C_{1/2}$.

2. Compare your answers to Key Question 11 with the graph in **Model 1**, and identify the order of the ozone decomposition reaction.

3. What is the rate law for the ozone decomposition reaction?

4. Write the integrated rate law for the ozone decomposition reaction in both exponential and logarithmic forms, and describe how they are related to the graph in **Model 1**.

5. Identify the order of the decomposition reaction with respect to ozone.

6. Identify the overall order of the ozone decomposition reaction.

7. Determine what would happen to the rate of decomposition if the concentration of ozone were doubled.

8. Determine the rate constant (magnitude and units) for the ozone decomposition reaction.

Got It!

1. Given the following rate law: Rate = k [CH$_3$Br] [OH$^-$]

    a. What is the reaction order with respect to bromomethane?

    b. What is the reaction order with respect to hydroxide?

    c. What is the overall reaction order?

    d. Identify what happens to the reaction rate if the concentration of bromomethane is cut in half.

    e. Identify what happens to the reaction rate if the concentration of hydroxide is increased by four times.

2. Label the following graphs as *zero order*, *first order*, or *second order* reactions.

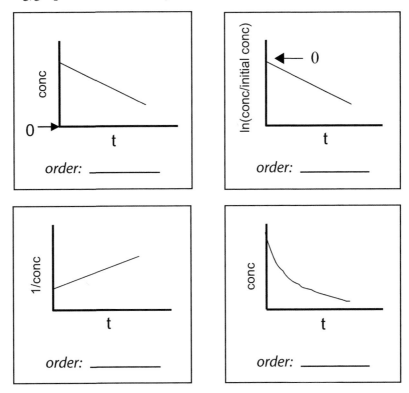

3. The statement, "The rate decreases as the reaction proceeds," does **not** apply to which one of the following?

    a. a zero order reaction   b. a first order reaction   c. a second order reaction

PROBLEMS

The data in the following table (molar concentration of N_2O_5 versus time) were obtained for the decomposition of dinitrogen pentoxide:

$$2\,N_2O_5(g) \rightarrow 4\,NO_2(g) + O_2(g)$$

time in s	0	1000	2000	8000	9000	15000	16000	20000	21000
conc in M	0.75	0.65	0.57	0.24	0.21	0.092	0.080	0.046	0.040

1. Determine—for example by making the appropriate graphs—whether this decomposition reaction is zero, first, or second order with respect to N_2O_5.

2. The stoichiometric coefficient for N_2O_5 in the reaction equation is 2. Depending on your answer to Problem 1, explain why the reaction order you obtained in Problem 1 is also 2, or explain why it may differ from 2.

3. Determine the rate constant for this reaction (magnitude and units).

4. Sketch a graph showing the relative concentrations of N_2O_5, NO_2, and O_2 as a function of time. Your graph should show these concentrations on the same scale, and should show the final concentrations relative to one another after equilibrium has been reached.

5. Write a paragraph explaining why your graph correctly describes how the concentrations vary with time.

13-2 Reaction Mechanisms

WHY?

A reaction mechanism is a sequence of elementary reactions that describes what chemists believe takes place as reactant molecules are transformed into product molecules. A rate law can be predicted from a proposed reaction mechanism, and you can compare predicted and experimental rate laws to determine whether a proposed reaction mechanism is consistent with experimental results. If a mechanism is correct, the predicted rate law must match the rate law that is determined experimentally, and the elementary reactions must be chemically reasonable.

LEARNING OBJECTIVES

- Understand how a series of elementary reactions describe an overall chemical reaction
- Develop the ability to derive a rate law from experimental data
- Develop the ability to derive a rate law from a proposed reaction mechanism

SUCCESS CRITERIA

- Determine the rate law for a reaction from experimental rate data
- For a given chemical reaction equation and experimentally determined rate law, design a suitable reaction mechanism

PREREQUISITE

13-1   *Rates of Chemical Reactions*

INFORMATION

A *chemical reaction mechanism* is a sequence of reactions that depict what is happening to reactants at the molecular level. These molecular-level reactions are called *elementary reactions*, and include simple processes like the transfer of an electron, the formation of a bond, the breaking of a bond, or the transfer of an atom from one reactant to another. The sum of the elementary reactions in a mechanism gives the overall balanced reaction equation.

A rate law can be predicted from a proposed reaction mechanism. If the predicted rate law does not match the experimental rate law, then the proposed mechanism cannot be correct. If the predicted rate law matches the experimental rate law, then it is possible that the proposed mechanism is correct. It is also possible for an incorrect mechanism to produce the correct rate law, so a match is not definitive! Often, methods other than kinetics, such as the detection of a reaction intermediate by an instrumental method, is required to provide further evidence for a mechanism.

An elementary reaction usually involves one or two molecules as reactants, and may involve collisions between molecules. Rarely are more than two molecules involved in an elementary reaction, since simultaneous collisions of three molecules are so unlikely. A *unimolecular* reaction involves only one molecule. A *bimolecular* reaction involves two molecules. The unlikely *termolecular* reaction involves the collision of three molecules.

MODEL 1	EXPERIMENTAL RATE LAW FOR THE OXIDATION OF NO

The rate law for the reaction of nitrogen monoxide with oxygen in the atmosphere has has been determined to be, from the following data collected at 15 °C,

$$\text{Rate} = -\frac{\Delta[O_2]}{\Delta t} = k[NO]^2[O_2]$$

[NO]$_0$ (mol / L)	[O$_2$]$_0$ (mol / L)	Initial Rate (mol / L min)
0.15	0.15	1.05
0.15	0.30	2.06
0.30	0.30	8.37

KEY QUESTIONS

1. Looking at the rate law in **Model 1**, what is:

   a. The order of the reaction with respect to oxygen?

   b. The order of the reaction with respect to nitrogen monoxide?

   c. The overall order of the reaction?

2. Based on the data in the table, what happens to the initial rate of reaction when the O$_2$ concentration is doubled while the NO concentration is kept constant? (Use quantitative descriptions in your answer, e.g., decreases by one half, increases by a factor of three, etc.)

3. Judging from the data in the table, what happens to the initial rate when the NO concentration is doubled while the O$_2$ concentration is kept constant?

4. Are your answers to Key Questions 2 and 3 consistent with the rate law in **Model 1** and the reaction orders with respect to NO and O_2? Describe how you determined whether they are consistent or inconsistent.

5. Based on the insight you gained by answering the Key Questions, how can data like that given in **Model 1** be used to determine an experimental rate law for a chemical reaction?

EXERCISES

1. Given the data in the table for **Model 1**, determine the value and units for the rate constant for the oxidation of nitrogen monoxide.

| MODEL 2 | REACTION MECHANISM |

The gas-phase reaction between nitrogen dioxide and carbon monoxide occurs in the two-step process below.

| Step 1 | $2 NO_2 \rightarrow NO_3 + NO$ | (slow) | bimolecular | rate $= k_1[NO_2]^2$ |
| Step 2 | $NO_3 + CO \rightarrow NO_2 + CO_2$ | (fast) | bimolecular | rate $= k_2[NO_3][CO]$ |

Overall Reaction: $NO_2 + CO \rightarrow NO + CO_2$ observed rate law: rate $= k[NO_2]^2$

Intermediate: NO_3

KEY QUESTIONS

6. How is the rate law for each elementary step determined?

7. Show that the overall reaction is consistent with the reaction mechanism.

8. The observed rate law does not show a dependence on the concentration of CO. What does this imply about its involvement in the reaction mechanism?

9. How is the overall rate law consistent with the mechanism?

MODEL 3	A MECHANISM FOR THE OXIDATION OF NO

A chemical reaction equation only summarizes the numbers of molecules or moles involved in the overall reaction. For example, the reaction equation for the oxidation of nitrogen monoxide,

$$2NO(g) + O_2(g) \rightarrow 2NO_2(g),$$

does not mean that two molecules of nitrogen monoxide collide with an oxygen molecule to produce two molecules of nitrogen dioxide. Such a collision between three molecules in the gas phase is extremely improbable, like three cars arriving at an intersection at exactly the same time from three different directions.

Chapter 13: Chemical Kinetics

The following mechanism has been proposed for this reaction, and it is composed of the following elementary reactions:

Step 1 (fast equilibrium): $2NO \underset{k_{-1}}{\overset{k_1}{\rightleftharpoons}} N_2O_2$

Step 2 (slow reaction): $N_2O_2 + O_2 \xrightarrow{k_2} 2NO_2$

If we add these two steps together, we obtain the overall reaction equation.

$$2\,NO + N_2O_2 + O_2 \rightarrow N_2O_2 + 2\,NO_2$$

$$2\,NO + O_2 \rightarrow 2\,NO_2$$

Step 1   involves two elementary reactions. The forward reaction is a bimolecular collision between 2 NO molecules to form N_2O_2. The reverse reaction is the dissociation of N_2O_2, which is a unimolecular reaction. The mechanism proposes that as result of these two reactions, an equilibrium concentration of N_2O_2 is established quickly.

Step 2   is a bimolecular reaction involving the collision of N_2O_2 with O_2. Since the rate of an elementary reaction is proportional to the concentration of the reactants, the rate for the reaction in Step 2 is:

(1)   Rate = $k_2[N_2O_2][O_2]$

The mechanism proposes that this rate is slow. The slow step in a reaction is called the *rate-limiting step*.

N_2O_2 is a reaction intermediate. A *reaction intermediate* is a chemical species that appears in an elementary reaction but does not appear in the overall reaction equation. To obtain the experimental rate law, the fast equilibrium in Step 1 is used to eliminate the reaction intermediate N_2O_2:

$$\text{rate forward} = \text{rate reverse}$$

$$k_1\left[NO\right]^2 = k_{-1}\left[N_2O_2\right]$$

$$\left[N_2O_2\right] = \frac{k_1}{k_{-1}}\left[NO\right]^2$$

$$Rate = \frac{k_2 k_1}{k_{-1}}\left[NO\right]^2\left[O_2\right]$$

The proposed mechanism therefore produces the experimentally observed rate law. Also, adding the elementary reactions produces the balanced reaction equation.

KEY QUESTIONS

10. How many elementary reactions are proposed in the mechanism for the oxidation of NO in **Model 3**?

11. Which of the elementary reactions in **Model 3** are unimolecular and which are bimolecular?

12. How is the rate law for an elementary reaction determined? For example, the rate for step 2 (k_2 [N_2O_2][O_2]) and the reverse reaction in step 1 of **Model 3** (k_{-1} [N_2O_2])?

13. In the proposed reaction mechanism, if the forward reaction in Step 1 were the slow step, what would be the rate law for the oxidation of NO?

14. An alternative mechanism for the oxidation of NO is an elementary reaction in which 3 molecules collide to give 2 molecules of NO_2.

$$2NO + O_2 \rightarrow 2NO_2$$

a. What is the rate law predicted by this reaction mechanism?

b. Is it consistent with the experimental rate law identified in **Model 1**?

c. Can the experimental rate law be used to identify whether this mechanism or the mechanism in **Model 1** is correct?

d. Why would you not expect a collision involving three molecules to be the mechanism for a reaction?

15. As a general rule, can the rate law be used to identify a unique, correct reaction mechanism, or can it only serve to eliminate incorrect mechanisms? Explain.

EXERCISES

2. Show that the sum of the two steps of the mechanism in **Model 3** equals the overall reaction equation.

3. Show that the experimental rate law for the oxidation of NO in **Model 1** can be obtained from the mechanism proposed in **Model 3**, and that the intermediate N_2O_2 does not appear in the rate law expression.

PROBLEMS

1. Develop a two-step mechanism for the oxidation of nitric oxide that involves oxygen atoms, and determine the rate law that it predicts. Note that oxygen atoms are very reactive and will therefore react quickly.

2. Use the kinetic data in the table below to determine the rate law for the following reaction:

$$aA + bB + cC \rightarrow dD + eE$$

Initial Concentrations (mol/L)			Initial Rate (mol/L s)
[A]	[B]	[C]	
0.40	0.15	0.40	3.20
0.15	0.30	0.30	1.60
0.60	0.15	0.10	0.75
0.20	0.15	0.10	0.25
0.15	0.30	0.15	0.40
0.15	0.15	0.15	0.40

3. The following mechanism is proposed for the reaction of chlorine with chloroform:

$$Cl_2 + CHCl_3 \rightarrow CCl_4 + HCl$$

a. Show that this mechanism produces the correct, balanced overall reaction equation.

b. Determine the rate law predicted by this mechanism.

ACTIVITY 13-3 Activation Energy and Catalysis

WHY?

Not all collisions between molecules result in a reaction. The molecules must have enough kinetic energy for the reaction to occur. The minimum energy that is needed for a reaction to occur is called the *activation energy*. *Catalysts* are substances that lower the activation energy of a reaction by providing an alternate reaction mechanism, thereby increasing the rate of the reaction. Catalysts are essential in the production of industrial chemicals. Biological catalysts, which are called *enzymes*, are essential for life and for the development of new pharmaceutical products.

LEARNING OBJECTIVES

- Understand the factors that limit the rate of a chemical reaction
- Be able to determine the activation energy of a chemical reaction from reaction rate data that depend on temperature
- Recognize how catalysts can increase reaction rates

SUCCESS CRITERIA

- Be able to produce a complete list of factors that affect the rate of a chemical reaction
- Identify three or more ways that the rate of a chemical reaction can be increased
- Correctly determine activation energies from reaction rate data
- Correctly predict how much a rate will increase due to a given change in activation energy

PREREQUISITES

10-3 *Kinetic Molecular Theory of Gases*

13-1 *Rates of Chemical Reactions*

13-2 *Reaction Mechanisms*

INFORMATION

For a reaction to occur, molecules must collide. The frequency of the collisions affects the rate of the reaction. The frequency can be changed by increasing or decreasing the concentrations of the reactants, or by changing the temperature, and therefore the kinetic energies, of the molecules.

But even when molecules collide, they do not all react. In order to react, the two molecules must be oriented in just the right way. Unfortunately, there is nothing that can be done to control molecular orientations, except in very sophisticated experiments.

Molecules must also have enough energy for the reaction to occur. That minimum energy is needed because existing bonds must be broken and new bonds must be formed. The point in the reaction at which this reconstruction occurs is called the *transition state* or *activated complex*, and the energy needed to reach the transition state is called the *activation energy*, E_a. Only those molecules that have an energy greater than or equal to the activation energy will react and produce products. The distribution of the energies of the molecules involved can be changed by raising or lowering the temperature. Remember that, for any given temperature, a Maxwell-Boltzmann distribution of molecular velocities exists, so only a fraction of the molecules present will have sufficient kinetic energy to react. However, at higher temperatures, a *greater* fraction of molecules will have sufficient kinetic energy.

MODEL 1   ENERGY VS REACTION COORDINATE DIAGRAM

An energy vs. reaction coordinate graph is used to show how the potential energy of the reactants changes as they become products. A reaction coordinate is a measure of the progress of the reaction along the reaction pathway.

Figure 1

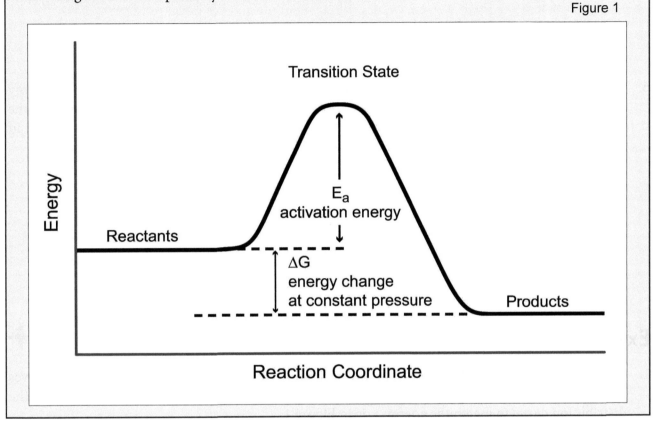

KEY QUESTIONS

1. Is the change in energy from reactants to products in **Model 1** positive or negative?

2. According to **Model 1**, which has the higher energy, the reactants or the transition state?

3. Which molecules are more likely to reach the transition state and react when they collide: those with high velocities and kinetic energies, or those with low velocities and kinetic energies?

4. As the temperature increases, does the fraction of molecules with high kinetic energies increase or decrease?

5. Given **Model 1** and your answers to Key Questions 1–4, why do you think the rate of a chemical reaction increases with increasing temperature?

6. Draw an arrow on the diagram in **Model 1** to indicate the magnitude of the activation energy for the reverse reaction (products going back to reactants). Do you think the rate constant of the reverse reaction will be larger or smaller than that of the forward reaction? Explain your answer in terms of the fraction of molecules that have enough kinetic energy to reach the transition state.

EXERCISES

1. Referring to Activity 10-3, if necessary, add to Figure 1 by sketching energy distribution curves for reactant molecules at two different temperatures. You will need to rotate the energy axis of the distribution curve to match the energy axis in Figure 1 .

2. Hydrogen and chlorine react to produce hydrochloric acid, but the reverse reaction also occurs at a slower rate. Consider the reaction $2HCl \rightarrow H_2 + Cl_2$. Draw two diagrams, one showing an orientation of the two HCl molecules that is unfavorable for this reaction, and one showing an orientation that is favorable for this reaction.

3. Give two or more reasons why some collisions between molecules might not result in a chemical reaction.

4. Draw an energy vs reaction coordinate diagram to illustrate a reaction in which the energy of the products is greater than the energy of the reactants. Label all quantities, as in **Model 1**.

4. Using your diagram from Exercise 4, identify which reaction (forward or reverse) has the larger activation energy and which has the larger rate constant.

INFORMATION

Because a reaction is faster at higher temperatures, the rate constant for the reaction must also be larger. The activation energy of a reaction can therefore be determined from experimental measurements of the rate constant at several temperatures. When the collected data are plotted in the form $\ln(k)$ vs $1/T$, as shown in Figure 2, a straight line is produced. This result means that the rate constant varies exponentially with $1/T$, as given by the Arrhenius equation.

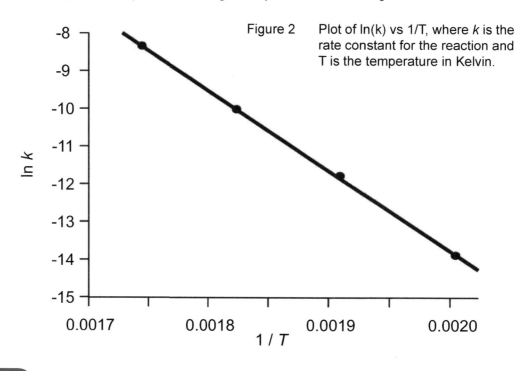

Figure 2   Plot of ln(k) vs 1/T, where k is the rate constant for the reaction and T is the temperature in Kelvin.

MODEL 2   THE ARRHENIUS EQUATION

$$k = Ae^{-\frac{E_a}{RT}}$$

The constant A is referred to as a *frequency factor*, and is determined by how often molecules collide when the concentrations are 1M, and whether the molecules are properly oriented when they collide. A is the value that the rate constant would have if all the molecules had enough energy to react, e.g., when the activation energy $E_a = 0$ or the Kelvin temperature T is very large.

KEY QUESTIONS

7.   Since the gas constant R has units of J/mol K, what are the units of E_a?

8. Does the Arrhenius equation predict that the rate constant will increase or decrease if the activation energy gets larger?

9. Does the Arrhenius equation predict that the rate constant will increase or decrease as the temperature increases?

EXERCISES

5. Show that the Arrhenius equation produces the following equation by taking the natural logarithm of both sides and using the property $\ln(ab) = \ln(a) + \ln(b)$.

$$\ln k = \ln A - \frac{E_a}{R}\frac{1}{T}$$

6. The equation in Exercise 5 creates a straight line when $\ln(k)$ is plotted versus $1/T$. Identify the quantities in the equation that determine the slope of the line and the quantity that determines the intercept at $1/T = 0$.

7. The line in Figure 2 is described by $\ln(k) = 28.5 - (2.1 \times 10^4\,\text{K})/T$.

   a. Determine the frequency factor and the activation energy for the reaction.

   b. The rate constant for a second order reaction has units of $\text{L mol}^{-1}\,\text{s}^{-1}$. What are the units of the frequency factor, A, for a second order reaction?

PROBLEMS

1. Fireflies flash at a rate that is temperature dependent. At 29 °C, the average firefly flashes at a rate of 3.3 flashes every 10 seconds. At 23 °C, the average rate is 2.7 flashes every 10 seconds. Use the Arrhenius equation to determine the activation energy for the flashing process.

INFORMATION

A catalyst, as shown in **Model 3**, changes the mechanism of a chemical reaction and lowers its activation energy. The catalyst participates in intermediate steps of the reaction, but it is neither produced nor consumed in the reaction, so the balanced reaction equation remains the same.

MODEL 3   CATALYSTS

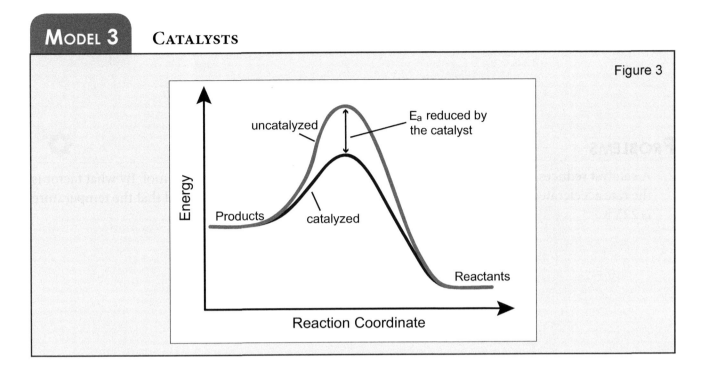

Figure 3

KEY QUESTIONS

10. What effect does a catalyst have on the activation energy of a reaction?

11. What effect does a catalyst have on the change in free energy of a reaction?

12. What effect does a catalyst have on the mechanism of a reaction?

13. What effect does a catalyst have on the stoichiometry of a reaction?

14. How does the rate of the rate limiting step in a reaction **with** a catalyst compare to its rate **without** the catalyst?

15. What are at least three ways that the rate of a chemical reaction can be increased?

PROBLEMS

2. A catalyst reduces the activation energy for a reaction from 20 kJ/mol to 4 kJ/mol. By what factor is the rate accelerated? Assume that the frequency factor A does not change, and that the temperature is 225 K.

14-1 Spontaneous Change and Entropy

WHY?

Sometimes, changes happen without outside intervention. These changes are called *spontaneous*. Spontaneous change occurs because things move from a less probable situation to a more probable one. The concept of *entropy* quantifies the probability of a situation and enables you to predict whether or not a change or a chemical reaction will be spontaneous. The Second Law of Thermodynamics summarizes this principle, by stating that the entropy of the universe increases during a spontaneous process. Entropy, unlike energy, is therefore *not* conserved; rather, it increases. These concepts are essential for predicting the spontaneity of chemical reactions and physical changes, or the temperature dependence of that spontaneity.

LEARNING OBJECTIVE

- Understand the meaning of spontaneous change and entropy

SUCCESS CRITERION

- Correctly identify whether entropy increases or decreases during a reaction

VOCABULARY

- *Spontaneous change* happens due to natural tendency, without an apparent external cause.

- *Entropy* is a measure of the dispersal of energy over the states available to a system.

INFORMATION

Energy does not naturally remain in one place; rather, it naturally tends to disperse. Entropy provides a quantitative measure of the extent to which energy has dispersed. Dispersal of energy occurs because there are more ways for the energy for the energy to spread out across many states than there are for it to be concentrated in one or a few states. You can see this dispersal occurring, for example, when a hot object that is surrounded by air, cools. Initially, the energy is contained in the hot object. As it cools, the energy is dispersed into the surrounding air molecules—the object cools down spontaneously, because there are more ways for the energy to be distributed across the surrounding air molecules than there are for it to be distributed within the object.

Entropy is related to the number of states that are accessible with a given amount of energy. The larger the number of states, the more dispersed the energy can be. A system that can be found in a large number of states often appears to be more disordered than a system that is found in only one state. Consequently entropy often is associated with the idea of order and disorder, but its real significance lies in its association with energy dispersal and accessible states.

MODEL 1   DISCOVER ENTROPY ON YOUR BOOKSHELF!

Entropy has a connection with the concept of probability. As an example of this connection, consider the number of possible arrangements for three books on a shelf, and let a particular arrangement of books be a "state." An arrangement set that can occur in many ways is more probable than an arrangement that can occur in only one way.

Arrangement Set	Constraint	Number of States
A	Three books are lined up alphabetically by author, starting on the left, with the binding facing outward and the titles on the binding facing left.	1
B	The books do not need to be arranged alphabetically by author, but the bindings face outward and the titles face left.	6
C	The books do not need to be arranged alphabetically by author; the bindings must face outward, but the titles can face right and left.	48
D	The books do not need to be arranged alphabetically by author; the bindings can face outward, inward, up, or down, and the titles can face right and left.	3072

KEY QUESTIONS

1. Which arrangement set has the maximum number of possible states?

2. Which arrangement set is most probable: A, B, C, or D?

3. Which arrangement set would you consider the most disordered?

4. Which arrangement set would you say has the highest entropy?

5. If you begin the semester with arrangement A, in which arrangement set will the state of your bookshelf be at the end of the semester, if you use your books and allow the order of your bookshelf to change spontaneously during the course of the semester (i.e., without extra care, effort, or intervention on your part)?

6. If spontaneous change is allowed to occur, would you expect the resulting state to be the most ordered arrangement? Explain in terms of the number of possible states and the probability of the most ordered arrangement.

7. Based on the model, what connections do you see between the number of accessible states, probability, entropy, and disorder?

8. Based on the model, what connection do you see between entropy and spontaneous change?

9. Identify another example of a spontaneous change that is accompanied by an increase in entropy.

INFORMATION

The following guidelines can be used to identify whether a physical or chemical change is accompanied by an increase or decrease in entropy. Entropy increases when particles can move around more freely and access a larger number of states. When a gas expands into a larger volume, the number of potential states (as specified by the position and momentum of each particle) available to the particles increases, as does the entropy. Consequently, the entropy increases when a chemical reaction produces a gas phase product from a solid or when a gas phase molecule dissociates into two or more particles. Similarly, when a solid melts or a liquid vaporizes, the particles have more positions available to them, and the entropy increases.

The decrease in molecular motion that accompanies precipitation would be expected to reduce the number of accessible states and decrease the entropy. Exceptions occur, however, because solvent molecules can bind tightly to ions when they form solutions. Consequently, the decreased entropy of the precipitating solute is counteracted by the solvent's larger increase in entropy. For example, the change in entropy for the precipitation of $AgCl$ is -33 J mol^{-1} K^{-1}, but for the precipitation of $MgCl_2$, it is $+115$ J mol^{-1} K^{-1}. There is no general rule for the entropy change of the dissolution or precipitation of a substance when both the solute and the solvent are taken into consideration.

MODEL 2   CHANGE IN ENTROPY OF A SYSTEM

When considering the spontaneity of a process, it is important to separate the entropy changes of the universe (ΔS_{univ}), surroundings (ΔS_{surr}), and system (ΔS_{sys}), since it is the positive entropy change of the universe that determines the overall spontaneity, according to the Second Law of Thermodynamics.

(Equation 1) $$\Delta S_{univ} = \Delta S_{surr} + \Delta S_{sys}$$

As you will see in the next activity, the entropy terms in Equation 1 can be redefined in terms of other thermodynamic parameters based on the system only, and the temperature dependence of spontaneity, ΔS_{univ}, will be quantifiable. For now, focusing on the entropy of the system, ΔS_{sys}, allows its contribution to the overall spontaneity of the process, ΔS_{univ}, to be somewhat predictable.

In general,

(Equation 2) $$\Delta S_{sys} = \Sigma m S_{products} - \Sigma n S_{reactants}$$

Unlike tabulated enthalpy of formation values that are defined for compounds as the enthalpy to form one mole of compound from its composite elements, ΔH_f°, tabulated entropy values are absolute values, listed as S instead of ΔS. The Third Law of Thermodynamics states that the entropy of a perfect crystal at absolute zero Kelvin is zero, so we deal with absolute values of entropy for substances. For the evaporation of 1 mole of water at 25 °C,

(Equation 3) $$\Delta S_{sys} = S_{vapor} - S_{liquid}$$

(Equation 4) $$\Delta S_{sys} = 188.84 \text{ J/(mol K)} - 69.95 \text{ J/(mol K)}$$

(Equation 5) $$\Delta S_{sys} = 118.89 \text{ J/(mol K)}$$

A positive value indicates that the change causes the entropy of the system to increase, contributing positively to the spontaneity of the process, since ΔS_{univ} must be positive.

EXERCISES

1. Identify whether the entropy of the system increases or decreases for each of the following changes and explain why.

   a. evaporation of a liquid

   b. freezing of a liquid

   c. precipitation of a solid from solution

   d. increase in the volume of a gas at constant temperature

   e. $2NaHCO_3(s) \rightarrow Na_2CO_3(s) + H_2O(g) + CO_2(g)$

   f. $2Fe(s) + 3 O_2(g) \rightarrow 2 Fe_2O_3(s)$

RESEARCH

Explain how the number of states for each book arrangement set in the model was determined.

Entropy of the Universe and Gibbs Free Energy

Why?

While the change in the entropy of a system can be used to determine its contribution to spontaneity of a process, one must consider the total change in entropy, i.e., the change in the entropy of the universe, to evaluate a process's overall spontaneity. This consideration leads to the concept of *free energy*, which is easier to apply because it deals only with the system being studied, rather than the entire universe. The concept of free energy is very useful. For example, you can use it to determine the concentrations of reactants and products in a chemical reaction at equilibrium, the voltage produced by batteries, and whether or not a chemical reaction will occur spontaneously.

Learning Objectives

- Understand the relationship between enthalpy and the entropy of a reaction's surroundings
- Understand the relationship between free energy and the entropy of the universe
- Understand the combined effects of enthalpy, entropy, and temperature when determining the spontaneity of a process.

Success Criteria

- Correctly predict increases and decreases in entropy and free energy
- Correctly identify spontaneous processes
- Accurately calculate changes in entropy and free energy

Prerequisites

06-2 *Internal Energy and Enthalpy*

14-1 *Spontaneous Change and Entropy*

Information

Free energy combines the ideas of enthalpy and entropy. Physically, the free energy change is the maximum work that can be done by a spontaneous process at constant temperature and pressure. If the process is not spontaneous, then the free energy is the minimum work needed to make the process occur.

Entropy, S, can be defined in terms of the number of states, W, that are accessible to a system with a given amount of energy:

$$S = k \ln(W) \quad \text{where } k = \text{Boltzmann's constant}$$

The change in entropy caused by a process, ΔS, can be defined in terms of the change in the number of accessible states, or in terms of the heat flow, q, into or out of a system at constant temperature:

$$\Delta S = q/T$$

The diagram in **Model 1** depicts heat flowing from a system into its surroundings. The sum of the system and its surroundings represents the universe. The change in the energy of the system is given by $\Delta E = q + w$, where q is the heat added to the system and w is the work done on the system. Since the heat is leaving the system in the model, it is given a negative sign, $-q_p$. The change in enthalpy of the system is also negative.

$$\Delta H = H_{final} - H_{initial} = -q_p$$

The subscript p means that the pressure is constant.

| MODEL | CHANGE IN THE ENTROPY OF THE UNIVERSE |

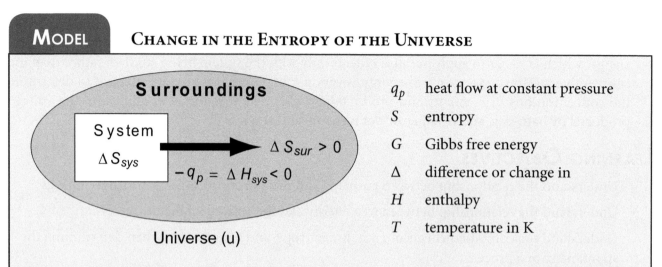

The entropy change of the universe is the total combined entropy change of the system and its surroundings.

(Equation 1) $\Delta S_u = \Delta S_{sur} + \Delta S_{sys}$

Heat flowing out of the system and into the surroundings increases the entropy of the surroundings. Since ΔH_{sys} is negative for an exothermic process, the following equation contains a minus sign, resulting in a positive ΔS_{sur} —an increase in entropy, as it should be.

(Equation 2) $\Delta S_{sur} = - \Delta H_{sys} / T$

Substituting equation (2) into equation (1) and multiplying through by $-T$ gives equation (3).

(Equation 3) $- T \Delta S_u = \Delta H_{sys} - T \Delta S_{sys}$

The quantity on the right side, which depends only on the system, is defined as the *free energy of the system*, or, more specifically, the *Gibbs free energy*.

(Equation 4) $\Delta G_{sys} = \Delta H_{sys} - T \Delta S_u$

The free energy is also related to the change in the entropy of the universe by equation (5)

(Equation 5) $\Delta G_{sys} = - T \Delta S_u$

The free energy accounts for the entropy of the universe in terms of the enthalpy and entropy of the system. These quantities (ΔG_{sys}, ΔH_{sys}, ΔS_{sys}) all refer to the system, so the subscript *sys* is not necessary and will no longer be used.

1. A process occurring in a system can cause the entropy of its surroundings to change. In the model, what is transferred from the system to cause the entropy of the surroundings to change?

2. When $\Delta H < 0$:

   a. does the energy of the system increase or decrease?

   b. does heat flow into or out of the surroundings?

   c. does the entropy of the surroundings increase or decrease?

3. According to the model, what do you need to know in order to calculate the change in the entropy of the surroundings?

4. What is the definition of the free energy change of the system in terms of the entropy of the universe?

5. For a process to be spontaneous, what must happen to the entropy of the universe?

6. For a process to be spontaneous, what must happen to the free energy of the system in which the process is occurring?

7. Why can a change be spontaneous even if the entropy of the system decreases?

8. What is an example of a spontaneous change in which the entropy of the system decreases?

EXERCISES

1. Show how the following key equation was derived in the model: $\Delta G = \Delta H - T\Delta S$

2. Calculate the change in entropy of the surroundings when 1.0 mol of methanol vaporizes at 64.7 °C. The heat of vaporization of methanol is 39.2 kJ/mol.

3. Calculate the free energy change of the following reaction, and predict whether it occurs spontaneously at 25 °C.

$$CH_4(g) + H_2O(g) \rightarrow CO(g) + 3H_2(g)$$

$$\Delta H = 206.1 \text{ kJ}, \quad \Delta S = 215 \text{ J/K}$$

4. The following table shows all the possible combinations of signs for the entropy and enthalpy changes in any process. Enter a +, –, or T in the ΔG column for each of these combinations. Use T for temperature if the sign of ΔG depends on temperature. If there is a temperature effect, specify whether the reaction is spontaneous at a high or a low temperature.

ΔS	ΔH	ΔG + or – or T	Spontaneous? enter yes or no or specify if it is spontaneous at a high T or low T
+	–		
–	+		
+	+		
–	–		

5. Using the information in Exercise 4, create a plot of ΔG vs. T, (in Kelvin) for the four combinations given in the table, assuming both ΔS and ΔH are **not** dependent on temperature.

6. Is an exothermic reaction or process always spontaneous? Explain.

7. Can an endothermic reaction or process be spontaneous? Explain.

8. Generally, when biological enzymes are heated, they no longer function as catalysts. This process is endothermic and spontaneous. Is the structure of the active enzyme more or less ordered than that of the inactive enzyme? Explain.

Dynamic Equilibrium and Le Châtelier's Principle

WHY?

When opposing forces or influences are balanced, a system is said to be at *equilibrium*. Equilibrium involving chemical reactions is always dynamic, because both forward and reverse reactions are continuously occurring. Equilibrium is reached when the rate of the reverse reaction equals the rate of the forward reaction. Le Châtelier's Principle provides a simple qualitative rule for analyzing the effects of disruptions to equilibrium situations. You can use this information to evaluate the extent to which a chemical reaction will proceed, to predict the potential effects of changes to an equilibrium, and to determine concentrations of reactants and products.

LEARNING OBJECTIVES

• Understand the meaning and implications of dynamic equilibrium

• Discover how systems will change in the process of reaching equilibrium

SUCCESS CRITERIA

• Correctly apply the idea of dynamic equilibrium to particular situations

• Correct predict how a system will change to reach equilibrium

PREREQUISITES

05-1 *Dissociation and Precipitation Reactions*

05-2 *Introduction to Acid – Base Reactions*

06-2 *Internal Energy and Enthalpy*

10-2 *Partial Pressure*

13-3 *Activation Energy and Catalysis*

MODEL 1   DYNAMIC EQUILIBRIUM

KEY QUESTIONS

1. Are the mice moving from habitat to habitat in the model?

2. Is the number of mice in each habitat changing?

3. In the same way that equilibrium is reached for the mice in the habitats, chemical equilibrium is reached in reactions when (select one):

    a. all reactions stop

    b. the forward reaction stops

    c. the concentrations of products and reactants become equal

    d. the rates of the forward and reverse reactions become equal

    e. the temperature becomes constant

4. Are the concentrations of A, B, C, and D changing if 100 molecules per second of A and B are being converted into C and D by the forward reaction, and 100 molecules per second of C and D are being converted into A and B by the reverse reaction? Explain.

5. What is an example of dynamic equilibrium that you have seen?

1. After the following reaction has reached equilibrium, all the carbon in CO is magically converted to the isotope carbon-14. Isotopes behave virtually identically chemically. Will all of the carbon-14 remain in the CO, or will some be found in the CO_2?

$$H_2O(g) + CO(g) \rightleftarrows H_2(g) + CO_2(g)$$

2. Consider the following reaction: $ClNO_2 + NO \rightleftarrows NO_2 + ClNO$

In one experiment, 1.0 mole $ClNO_2$ and 1.0 mol NO are put into a flask and heated to 350 °C. In a second experiment, 1.0 mol NO_2 and ClNO are put into another flask with the same volume as the first and heated to 350 °C. After equilibrium is reached in both cases, is there any difference in the composition of the mixtures in the two flasks? Explain.

MODEL 2	LE CHÂTELIER'S PRINCIPLE

> If a change is imposed on a system, the position of equilibrium will shift in a direction that tends to **reduce** the effect of that change.

Changes imposed on an equilibrium, and the response they produce, can be visualized using arrows indicating an increase (up) or a decrease (down) in concentration or temperature. For a generic system at equilibrium in which R_n are reactants and P_n are products, the equilibrium can be represented with the equation:

$$R_1 + R_2 \rightleftarrows P_1 + P_2$$

If R_1 is added to the equilibrium mixture, an up arrow can be used to represent this increased concentration. The equilibrium response is to decrease the concentration of R_1, represented by a down arrow. Since R_1 reacts with R_2 to form products, the concentration of R_2 decreases as a result, but the concentrations of P_1 and P_2 increase to re-establish the equilibrium position.

KEY QUESTIONS

6. What will happen to the number of mice in the two habitats in **Model 1** if the temperature in Habitat A rises because its air conditioning system fails? Explain your answer in terms of Le Châtelier's Principle. Assume that Habitat A becomes uncomfortably warm for mice, but Habitat B remains comfortable.

7. According to Le Châtelier's Principle, what will happen to a chemical reaction at equilibrium if a reaction product is removed?

8. According to Le Châtelier's Principle, what will happen to a chemical reaction at equilibrium if more reactant is added?

9. What is an example of Le Châtelier's Principle, applied in an area other than chemistry?

EXERCISES

3. An important endothermic gas phase reaction used in the commercial production of hydrogen is:

$$CH_4(g) + H_2O(g) \rightleftarrows 3 H_2(g) + CO(g)$$

Use Le Châtelier's Principle to explain what will happen to the number of moles of hydrogen present at equilibrium when...

a. carbon monoxide is removed.

b. water vapor is removed.

c. methane is added.

d. carbon monoxide is added.

e. argon gas is added to increase the total pressure, but the partial pressures of the reactants and products do not change.

f. total pressure is decreased by opening a valve to a second reaction vessel, which lowers the partial pressures of the reactants and products.

g. a catalyst is added. Catalysts change the mechanism of a reaction.

h. the temperature is increased.

PROBLEMS

1. Nitric oxide, a significant air pollutant, is formed from nitrogen and oxygen at high temperatures, e.g., in an automobile engine, by the following reaction:

$$N_2(g) + O_2(g) \rightleftarrows 2\,NO(g)$$

As the piston compresses these gases in the cylinder of an automobile engine, what happens to the equilibrium amount of NO in the cylinder? The movement of the piston both increases the pressure on the gases and decrease the volume of the container. Does the amount of NO in the cylinder increase, decrease, or stay the same? Explain why.

2. The following reaction is endothermic: $H_2(g) + I_2(g) \rightleftarrows 2HI(g)$

To obtain the highest yield of hydrogen iodide at equilibrium, what combination of temperature and pressure would you use? Explain.

a. high T and P   b. low T and P   c. high T, low P   d. low T, high P

The Reaction Quotient & Equilibrium Constant

WHY?

All chemical reactions (and many other processes as well) eventually reach an equilibrium, at which point changes in concentrations of products and reactants no longer take place. For chemical reactions, this point is described by a characteristic constant called the *equilibrium constant*: a mathematical expression of the ratio of product and reactant concentrations at a specific temperature. You can use the equilibrium constant to predict the amount of certain reactants that will become products, and to calculate the equilibrium concentrations of reactants and products under various conditions. The *reaction quotient* is similar to the equilibrium constant in function, but it describes the system at any point before equilibrium is established, at any arbitrary combination of reactants and products.

LEARNING OBJECTIVES

- Learn how to use the reaction quotient and equilibrium constant expressions
- Understand how the reaction quotient can be used to predict the direction of a chemical reaction

SUCCESS CRITERIA

- Accurately calculate values for reaction quotients and equilibrium constants
- Correctly predict the direction a reaction will proceed in order to reach equilibrium

PREREQUISITES

04-1   *Balanced Chemical Reaction Equations*

10-1   *The Ideal Gas Law*

10-2   *Partial Pressure*

15-1   *Dynamic Equilibrium and Le Châtelier's Principle*

MODEL   THE REACTION QUOTIENT AND EQUILIBRIUM CONSTANT

Reaction equations that are written with a single arrow proceeding from products to reactants imply that only reactants exist before the reaction takes place, and only products exist after the reaction has proceeded to completion. This is a simplification. A more accurate reaction equation uses the double arrow to show that both reactants and products are present before and after equilibrium is established.

For a general balanced reaction equation $a\,A + b\,B \rightleftarrows c\,C + d\,D$

the Reaction Quotient Q is written as: $Q = \dfrac{[C]^c [D]^d}{[A]^a [B]^b}$

where the quantities in square brackets are the molar concentrations divided by the standard concentration of 1M, so the units cancel. The concentrations of the products are in the numerator and the concentrations of the reactants are in the denominator, and each is raised to the power given by its stoichiometric coefficient.

When the concentrations at equilibrium are used, the Reaction Quotient becomes the Equilibrium Constant K or K_c (the subscript **c** denoting concentration). The Equilibrium Constant characterizes a reaction that is at equilibrium at a given temperature and is given by:

$$K = \frac{[C]_{eq}^{c}[D]_{eq}^{d}}{[A]_{eq}^{a}[B]_{eq}^{b}}$$

KEY QUESTIONS

1. What items go into the numerator in the reaction quotient and equilibrium constant expressions?

2. What items go into the denominator in the reaction quotient and equilibrium constant expressions?

3. What determines the exponent for each item in the reaction quotient and equilibrium constant expressions?

4. Under the following conditions, at equilibrium, will reactants or products be present in the largest amounts?

   a. When the equilibrium constant is small.

   b. When the equilibrium constant is large.

5. If Q is larger than K, will the reaction proceed in the forward direction or the reverse direction to reach equilibrium? Explain your reasoning.

6. If Q is smaller than K, will the reaction proceed in the forward direction or the reverse direction to reach equilibrium? Explain your reasoning.

7. When will $Q = K$?

INFORMATION

For gas phase equilibrium reactions, it is also possible to express the equilibrium quotient, Q_p, and equilibrium constant, K_p, in terms of partial pressures of the reactants and products. The only situation in which $K_p = K_c$ for a given reaction at a given temperature when it is a gas phase reaction in which the overall number of moles of reactant and product gases does not change ($\Delta n = 0$).

For any other gas phase equilibrium in which $\Delta n \neq 0$: $\boldsymbol{K_p = K_c(RT)^{\Delta n}}$

where R = 0.08206 in units (L $\cdot$ atm/mol $\cdot$ K), T is in Kelvin, concentration is in M, and P is in atm, and Δn = (moles of product gas – moles of reactant gas).

EXERCISES

1. What is the equilibrium constant for the following reaction?

$$2\,SO_2\,(g) + O_2\,(g) \rightleftarrows 2\,SO_3\,(g)$$

2. Write the equilibrium constant expression, K_c, for the following reaction. Note that the concentration of a pure solid is a constant. The moles per liter do not change during the reaction; the solid is just consumed. Consequently, the concentration of the solid is not included in the equilibrium constant expression.

$$H_2O\,(g) + C\,(s) \rightleftarrows H_2\,(g) + CO\,(g)$$

3. For reactions in solution, molar concentrations are usually used in equilibrium constant expressions (designated by K or K_c). In gases, partial pressures can also be used (designated by K_p). Equilibrium partial pressures of NOCl, NO, and Cl_2 in a container at 300 K are 1.2 atm, 0.050 atm, and 0.30 atm, respectively. Calculate a value for the equilibrium constant K_p for the following reaction, then calculate the corresponding value for K_c for the same reaction.

$$2 \, NO(g) + Cl_2 \, (g) \rightleftarrows 2 \, NOCl(g)$$

Got It!

1. Hydrogen iodide decomposes to give hydrogen and iodine. The equilibrium constant is about 2.0×10^{-3} at some constant temperature. If you were to place some hydrogen iodide in a container and seal it, what distribution of elements would you expect to find at equilibrium: mostly hydrogen iodide or mostly hydrogen and iodine? Explain how the value of the equilibrium constant helped you answer this question.

WHY?

The change in free energy for a chemical reaction depends on the temperature at which the reaction takes place and the concentrations of the reactants and the products. Like chemists in research and manufacturing, you too can use free energy to identify suitable temperatures for reactions, to determine the direction a reaction will proceed to reach equilibrium, and to obtain values for equilibrium constants and equilibrium concentrations.

LEARNING OBJECTIVES

- Understand the relationship between free energy and the reaction quotient expression
- Identify the effect of temperature on a reaction's change in free energy and its equilibrium constants

SUCCESS CRITERIA

- Calculate the free energy change for a reaction at various temperatures and various concentrations of reactants and products
- Identify the direction in which a reaction will spontaneously proceed
- Calculate one or more of $\Delta G°$, $\Delta H°$, $\Delta S°$; or, given these values, calculate K

PREREQUISITES

06-3 *Hess's Law: Enthalpy is a State Function*

14-2 *Entropy of the Universe and Gibbs Free Energy*

15-2 *The Reaction Quotient & Equilibrium Constant*

INFORMATION

The change in free energy of a chemical reaction under standard conditions is called the *standard free energy change*, $\Delta G°$. Conditions are *standard* when the pressure is 1 atm, and all reactants and products are either pure solids, liquids, or gases, or are solutes at 1 M concentration.

The free energy change in a chemical reaction (ΔG_T) at some Kelvin temperature T, is related by the following equation to the standard free energy change at that temperature, $\Delta G_T°$, and the reaction quotient, Q, which involves the concentrations of reactants and products. (R is the ideal gas constant.)

$$\Delta G_T = \Delta G_T° + RT \ln(Q)$$

When reactants and products are in their standard states, then Q is 1, because all concentrations are 1 M. Then $\ln(Q) = \ln(1) = 0$, $\Delta G_T = \Delta G_T°$, and the change in the free energy is just the standard free energy change.

An increase in the entropy of the universe or a decrease in the free energy of the system is the driving force behind spontaneous change. Concentrations in a chemical reaction change spontaneously in ways that lower the free energy. Equilibrium is reached in a chemical reaction when the total free energy can no longer decrease because the free energy of the reactants is equal to the free energy of the products. The free energy change for the reaction, ΔG_T, is then equal to 0.

Since $Q = K$ at equilibrium, $\Delta G_T = \Delta G_T^\circ + RT \ln(K) = 0$

which means that: $\Delta G_T^\circ = -RT \ln(K)$

which can be rearranged to
provide an expression for K: $K = e^{-\frac{\Delta G^\circ}{RT}}$

and also, since $\Delta G^\circ = \Delta H^\circ - T\Delta S^\circ$, $\Delta H^\circ - T\Delta S^\circ = -RT \ln(K)$

and after dividing by $-RT$, $\ln(K) = -\Delta H^\circ / RT + \Delta S^\circ / R$

A plot of $\ln(K)$ vs. $1/T$ allows determination of both ΔH° and ΔS° from experimental data.

KEY QUESTIONS

1. According to the information provided, what is the relationship between the free energy change for a chemical reaction and the concentrations of the reaction's reactants and products?

2. What makes the free energy change (ΔG) for a chemical reaction differ from its standard free energy change (ΔG°)?

3. According to the information provided, what is the free energy change for a chemical reaction (or any other process) at equilibrium?

4. Given two reactions with a positive standard free energy change, will the reaction with the smaller change have the larger or smaller equilibrium constant? Explain why your answer is reasonable.

5. According to the information provided, why are equilibrium constants for a chemical reaction temperature dependent?

EXERCISES

1. Using the equation $\Delta G_T = \Delta G_T^\circ + RT \ln(Q)$, show that $\Delta G_T^\circ = -RT \ln(K)$ when reactants and products in a chemical reaction are present at their equilibrium concentrations.

2. Use the equations in the information section to show that the free energy change for a chemical reaction when all concentrations are 1M is equal to the standard free energy change for that reaction.

3. The change in standard free energy for the synthesis of water from hydrogen and oxygen gases is -237 kJ/mol. Write the balanced reaction equation and the equilibrium constant expression for this reaction, and determine the value of the equilibrium constant at 25 °C.

4. Calculate the equilibrium constant for the synthesis of ammonia at 350 °C, given that $\Delta H° = -91.8$ kJ mol^{-1} and $\Delta S° = -198.1$ J mol^{-1} K^{-1}.

$$N_2(g) + 3H_2(g) \rightleftarrows 2NH_3(g)$$

5. Given that $\Delta G° = 2.8$ kJ mol^{-1}, calculate the free energy change for the decomposition of dinitrogen tetroxide at 298 K when $P_{N_2O_4} = 5$ atm and $P_{NO_2} = 2$ atm

$$N_2O_4(g) \rightleftarrows 2NO_2(g)$$

6. Calculate the equilibrium constant for the reaction of carbon monoxide and hydrogen at 298 K, $CO(g) + 2H_2(g) \rightleftarrows CH_3OH(l)$, given the following information:

$$\Delta G_f° \, CO(g) = -137.2 \text{ kJ mol}^{-1}$$

$$\Delta G_f° \, H_2(g) = 0 \text{ kJ mol}^{-1}$$

$$\Delta G_f° \, CH_3OH(l) = -166.6 \text{ kJ mol}^{-1}$$

7. The auto-ionization of water, which will be discussed in much more detail later, is described by the following equilibrium:

$$2 H_2O(l) \rightleftarrows H_3O^+(aq) + OH^-(aq)$$

This equilibrium reaction is essential to all aqueous reaction chemistry, and is given a special name for its equilibrium constant, K_w. The value of K_w varies with temperature, as shown in the table of data at right. Using this data, determine the values of $\Delta H°$ and $\Delta S°$ for this equilibrium. Use of a spreadsheet program like Microsoft Excel is recommended for this problem. Create an appropriate plot and use correct data analysis in your procedure.

T, °C	K_w
0	1.12×10^{-15}
25	1.02×10^{-14}
50	5.49×10^{-14}
75	1.99×10^{-13}
100	5.62×10^{-13}

8. Which graph below best describes the variation of $\Delta G°$ with temperature $y = \Delta G°$ and $x = T$, and both $\Delta H°$ and $\Delta S°$ are positive.

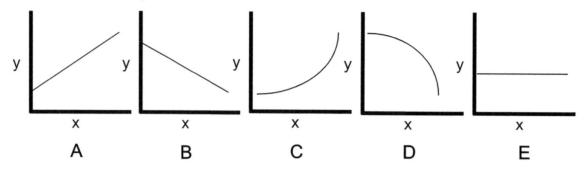

9. Which graph below best describes the variation of $\ln(K)$ with temperature $y = \ln(K)$ and $x = 1/T$, and both $\Delta H°$ and $\Delta S°$ are negative. Explain your reasoning.

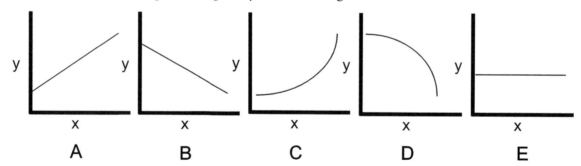

ACTIVITY 15-4 Solving Equilibrium Problems: The RICE Table Methodology

WHY?

When trying to solve problems, especially equilibrium problems, it is often helpful to use a strategy or a *methodology*: a logical sequence of steps. Following the steps of a methodology will help you achieve success as you complete homework assignments, take examinations, and solve problems throughout your life.

LEARNING OBJECTIVE

- Master a methodology for solving general chemistry equilibrium problems

SUCCESS CRITERIA

- Successfully apply the RICE table methodology to equilibrium problems
- Accurately solve problems involving reactant and product concentrations and equilibrium constants

PREREQUISITES

04-4 *Solving Solution Stoichiometry Problems*

15-2 *The Reaction Quotient & Equilibrium Constant*

INFORMATION

In your previous study of chemical reactions that were assumed to proceed to completion (Activity 04-4), you may have mastered the RICE table method for limiting reactant problems. These tables—sometimes referred to as *reaction tables*—have some advantages over other methods of solving limiting reactant problems, because they can be used to simultaneously find a limiting reactant, calculate how much of a product is produced, and calculate how much excess reactant remains. For RICE tables it was important to begin by converting all reactant quantities to moles or mmoles, and to understand that all results were in moles or mmoles as well. If you have not mastered that method, return to Activity 04-4 **Model 2** before proceeding with this new activity. These two applications of RICE tables differ in several ways. In the following model, quantities are expressed in terms of concentration, rather than moles, and terms describing equilibrium concentrations are not set to zero; rather, these expressions are used to determine x values for use in equilibrium calculation

MODEL	**THE RICE TABLE METHODOLOGY** **APPLIED TO EQUILIBRIUM PROBLEMS**

Initially 1.50 moles of $N_2(g)$ and 3.50 moles of $H_2(g)$ were added to a 1.00 L container at 700 °C. When the reaction,

$$3H_2(g) + N_2(g) \rightleftarrows 2NH_3(g)$$

reached equilibrium, the concentration of $NH_3(g)$ was 0.540 M. What is the value of the equilibrium constant for this reaction at 700 °C?

The RICE Table Methodology

Steps in the Strategy	Example
Before using the table for equilibrium problems, convert all quantities to concentration in M for K_c or atmospheres for K_p.	1.50 mol N_2 /1.00 L = 1.50 M N_2 3.50 mol H_2 /1.00 L = 3.50 M H_2 0 M NH_3
Step R: Write the **Reaction equation**.	3 $H_2(g)$ + $N_2(g)$ $\rightleftarrows$ $2NH_3(g)$
Step I: List the **Initial** concentrations (or partial pressures) of reactants and products.	3.5 mol 1.5 mol 0 mol
Step C: Using the stoichiometric coefficients in the reaction equation, write the **Change** in the concentration of each substance due to the reaction.	$-3x$ $-x$ $+2x$
Step E: Write the amounts of substances present after the reaction reaches **Equilibrium** or goes to completion.	$3.5-3x$ mol $1.5-x$ mol $2x$ mol

Finally, solve for unknowns as requested by the problem statement.

The problem statement gives the concentration of NH_3 at equilibrium, so

$$2x \text{ mol} / 1L = 0.540 \text{ M} \qquad x = 0.270 \text{ mol}$$

So the equilibrium concentrations are

$(3.5 - (3 \times 0.270))$ mol/L	$(1.5 - 0.270)$ mol/L	(2×0.270) mol/L
2.69 M	1.23 M	0.540 M

$$K = \frac{[NH_3]^2}{[H_2]^3[N_2]} \quad \text{is the equilibrium constant.}$$

Substituting into the expression for the equilibrium constant produces: $\quad K = \dfrac{(0.540)^2}{(2.69)^3(1.23)} = 0.0122$

1. In the RICE table methodology, what do the letters R, I, C, and E represent?

2. Why is writing the balanced reaction equation an important part of the methodology?

3. Are there any steps in the methodology that you feel are not useful? Why?

4. Can you improve the methodology by altering the steps or changing the order of steps? Explain.

5. What insight about solving equilibrium problems did your team gain by examining the methodology?

EXERCISE

1. Initially, 1.0 mole of NO and 1.0 mol of Cl_2 were added to a 1.0 L container. When the reaction,

$$2NO(g) + Cl_2\ (g) \leftrightarrows 2NOCl\ (g),$$

reached equilibrium, the concentration of NOCl was 0.96 M. Using the RICE table methodology, complete the table below and determine the value of the equilibrium constant (K_c) for this reaction.

The RICE Table Methodology

Steps in the Strategy	
Step R: Write the **Reaction equation**.	
Step I: List the **Initial** concentrations (or partial pressures) of reactants and products.	
Step C: Using the stoichiometric coefficients in the reaction equation, write the **Change** in the concentration of each substance due to the reaction.	
Step E: Write the amounts of substances present after the reaction reaches **Equilibrium** or goes to completion.	
Then solve for unknowns as requested by the problem statement.	

PROBLEMS

1. If a 3.00 L flask at 187 K is filled with 6.0 moles of NOCl (g), what are the equilibrium concentrations of the three gases? 33% of NOCl is dissociated at equilibrium. What is the equilibrium constant K_c? The relevant reaction is

$$2\ NOCl\ (g) \rightleftarrows 2\ NO(g) + Cl_2\ (g)$$

Foundations of Chemistry: 5th ed.

2. Nitrogen dioxide is introduced into a flask at a pressure of one atmosphere (1.00 atm). After some time it dimerizes to produce N_2O_4.

   a. What is the equilibrium partial pressure of the N_2O_4 at a temperature where the equilibrium constant for dimerization is $K_p = 3.33$?

   b. Calculate the total pressure in the flask.

   c. Explain why the total pressure is more than, less than, or still equal to one atmosphere.

                                                                              Chapter 15: Chemical Equilibrium

3. Gaseous carbon dioxide is partially decomposed according to the following equation. 1.00 atm of CO_2 is placed in a closed container at 2500 K, and 2.1% of the molecules decompose. Determine the equilibrium constant K_p at this temperature.

$$2CO_2(g) \leftrightarrows 2CO(g) + O_2(g)$$

4. A sample of solid ammonium nitrate was placed in an evacuated container and then heated so it decomposed explosively according to the following equation. At equilibrium, the total pressure in the container was found to be 5.20 atm at 650 °C. Determine K_p for this reaction.

$$NH_4NO_3(s) \leftrightarrows N_2O(g) + 2H_2O(g)$$

ACTIVITY 16-1 The pH Scale and Water Autoionization

WHY?

Water and aqueous reaction chemistry are essential to all biological and environmental systems, and therefore to our everyday life. The acidity or basicity of a solution determines the behavior of many other reactions as well. In aqueous solution, the *active agents* often include the hydronium ion, H_3O^+, which is created by adding a proton to a water molecule, or the hydroxide ion, OH^-, which is produced when a proton is removed from a water molecule. In all aqueous solutions, these ions are produced by an equilibrium reaction. In all aqueous solutions, these ions are produced by an equilibrium reaction. The concentrations of these two species can vary significantly across solutions, so it is convenient to use the base-ten *logarithmic scale* to quantify concentrations, using a number between 0 and 14. You may recognize these numbers from the pH scale (or the less common pOH scale). You need to understand and be able to use these scales in order to measure and control acidity and to fully utilize acid-base reactions.

LEARNING OBJECTIVES

- Understand and use the pH scale
- Be able to generalize a logarithmic scale for any quantity; specifically, the concentration of the hydroxide ion (pOH)
- Understand how the pH of pure water is determined by auto-ionization

SUCCESS CRITERION

- Correctly compare and convert values between the pH and pOH scales and hydronium and hydroxide ion concentrations

PREREQUISITES

03-1   *Nomenclature: Naming Compounds*

05-2   *Introduction to Acid – Base Reactions*

15-1   *Dynamic Equilibrium and Le Châtelier's Principle*

15-2   *The Reaction Quotient & Equilibrium Constant*

INFORMATION

At 25 °C, the concentration of hydronium ion in pure water is 1.0×10^{-7} M, and the hydroxide ion concentration is also 1.0×10^{-7} M, due to the auto-ionization equilibrium reaction shown in equation 1.

(Equation 1) $H_2O(l) + H_2O(l) \leftrightarrows H_3O^+ (aq) + OH^-(aq)$

Since the concentrations of hydronium ion and hydroxide ion are equal, pure water is said to be neutral. Because the liquid water is not shown in the equilibrium expression for this reaction, the equilibrium expression contains only ionic concentrations, and the equilibrium constant has a special name, K_w, as shown in Equation 2.

(Equation 2) $K_w = [H_3O^+][OH^-] = (10^{-7})(10^{-7}) = 1.0 \times 10^{-14}$

If acids or bases are added to water, the hydronium and hydroxide ion concentrations change. If the hydronium ion concentration increases, the solution becomes acidic. If the hydroxide ion concentration increases, the solution becomes basic.

MODEL 1   LOGARITHMS AND pH

Properties of Logarithms

$\log(100) = 2$ since $100 = 10^2$

$\log(0.01) = -2$ since $0.01 = 10^{-2}$

$\log(1) = 0$ since $1 = 10^0$

$\log(0.00851) = -2.07$ since $0.00851 = 10^{-2.07}$

$\log(1.122 \times 10^{-14}) = -13.95$ since $1.122 \times 10^{-14} = 10^{-13.95}$

Definition of pH

$pH = -\log[H_3O^+]$

$[H_3O^+] = 0.01\ M = 10^{-2}$ so $pH = 2$

$[H_3O^+] = 0.00851\ M = 10^{-2.07}$ so $pH = 2.07$

$[H_3O^+] = 1.122 \times 10^{-14}\ M = 10^{-13.95}$ so $pH = 13.95$

KEY QUESTIONS

1. According to **Model 1**, it is necessary to use base 10 logarithms, symbolized with *log*, rather than the natural logarithm (base *e*), which is written *ln*. Show the difference in results by using your calculator to find the value of $\log(1 \times 10^{-7})$ and $\ln(1 \times 10^{-7})$.

2 According to **Model 1**, how many significant figures are used in the logarithm of 0.00851?

3. According to the model, if you are given the logarithm of a number *x*, what do you need to do to obtain *x* itself?

4. According to **Model 1**, if you are given the pH of a solution, what do you need to do to obtain the hydronium ion concentration?

5. According to **Model 1**, if you are given the hydronium ion concentration of a solution, what do you need to do to obtain its pH?

6. The pH scale ranges from 0 to 14. What is the range of hydronium ion concentration covered by this scale?

7. Why is the pH scale used to measure hydronium ion concentrations?

EXERCISES

1. Using your calculator, check the mathematical calculations shown in **Model 1** to make sure that they are correct, and confirm that you know how to convert between numbers and their logarithms. Perform the following calculations:

   a. $10^3 =$

   b. $\log(0.001) =$

   c. $10^{2.07} =$

   d. $\log(1.22 \times 10^{-13}) =$

2. Based on the information in **Model 1**, define *pH*, using words.

CALCULATIONS INVOLVING THE WATER AUTO-IONIZATION EQUILIBRIUM

Equations 1 and 2 in the Information section describe the auto-ionization equilibrium of water at 25 °C. In neutral solution, $[H_3O^+] = [OH^-] = 1.0 \times 10^{-7}$ M, so $K_w = 1.0 \times 10^{-14}$.

In an acidic solution, where pH is lower than 7.0, the $[OH^-]$ must be proportionally lower than 1.0×10^{-7} in order to maintain their product as $K_w = 1.0 \times 10^{-14}$.

If $[H_3O^+] = 1.0 \times 10^{-4}$ M, then $[OH^-]$ can be found by rearranging the K_w expression, as in equation 3.

$$(\text{Equation 3}) \qquad \frac{K_w}{[H_3O^+]} = [OH^-] \quad \text{so} \quad \frac{\left(1.0 \times 10^{-14}\right)}{\left(1.0 \times 10^{-4}\right)} = [OH^-] = 1.0 \times 10^{-10}\,M$$

KEY QUESTIONS

8. What is the pH of a neutral aqueous solution at 25 °C, given that the hydronium ion concentration is 1.0×10^{-7} M?

9. Does a pH of 2.7 describe a solution with a higher or lower hydronium ion concentration, as compared to a neutral solution? Is such a solution called *acidic* or *basic*?

10. Compared to a neutral solution, would you expect a basic solution to have a higher or lower pH? A higher or lower hydronium ion concentration?

11. How could you define a quantity labeled "pOH" so that it would be analogous to the definition of pH?

INFORMATION

For any quantity, a *p-scale* can be defined as the negative of the logarithm of that quantity:

$$pX = -\log x \text{ so } pOH = -\log[OH^-]$$

EXERCISES

3. Identify which of the following substances has the highest hydronium ion concentration:

    chocolate milk: $[H_3O^+] = 9.77 \times 10^{-7}$ M

    orange juice: $[H_3O^+] = 1.99 \times 10^{-4}$ M

    toothpaste: $[H_3O^+] = 1.25 \times 10^{-10}$ M

4. Order these substances from most acidic to most basic: chocolate milk, orange juice, toothpaste.

5. Which of these substances has the highest pH value: milk, pickle juice, or oven cleaner?

6. Convert the following pH values into hydronium ion concentrations.

pH	1.0	5.3	7.0	13.3
Calculated [H$_3$O$^+$]				

7. Convert the following hydroxide ion concentrations into pOH values, first by estimating, and then by using your calculator. To estimate, only consider the exponent (power of ten).

[OH$^-$]	1 × 10^{-13}	2 × 10^{-9}	1 × 10^{-7}	0.2
Estimated pOH				
Calculated pOH				

8. The values in corresponding columns of the tables in Exercises 6 and 7 were obtained for the same solutions. All the solutions were at 25 °C. For each column in the tables in Exercises 6 and 7, add the pH and calculated pOH values, and put your result in the table below. The first column has been done for you.

(pH + calculated pOH)	14.0			

9. For each column in the tables in Exercises 6 and 7, determine the product $[H_3O^+] \times [OH^-]$, and put your result in the table below. The first column has been done for you.

(calculated $[H_3O^+] \times [OH^-]$)	1.0×10^{-14}			

10. Write a general statement describing your discovery regarding the sum of the pH and pOH values for an aqueous solution at 25 °C.

11. Write a general statement describing your discovery regarding the product of hydronium and hydroxide ion concentrations for an aqueous solution at 25 °C.

Got It!

1. On the diagram below, show how it is possible to reach any corner of the square from any other corner, for a given solution at 25 °C, using relationships you have encountered in this activity.

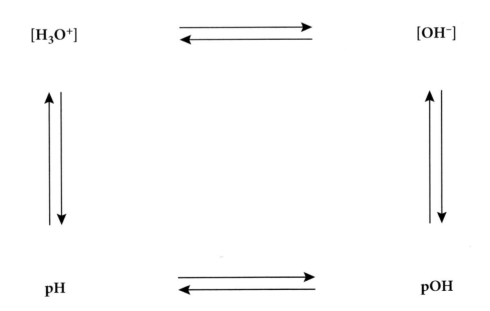

$[H_3O^+]$ ⟶⟵ $[OH^-]$

pH ⟶⟵ pOH

2. Complete the following statements:

    a. As hydronium ion concentration increases, pH

        ☐ *increases*   ☐ *decreases*

    b. As hydronium ion concentration increases, hydroxide ion concentration

        ☐ *increases*   ☐ *decreases*

    c. As pH decreases, pOH

        ☐ *increases*   ☐ *decreases*

    d. As pH decreases, hydronium ion concentration

        ☐ *increases*   ☐ *decreases*

3. Sports drinks have a pH of 3.05.

    a. What is their hydronium ion concentration?

    b. Are these commercial product neutral, basic, or acidic?

    c. What is their pOH?

    d. What is its hydroxide ion concentration?

4. As more water is added to an aqueous solution of a strong acid (with pH = 3 at 25 °C), the pH (select one):

    a. increases until a constant value of 7 is reached.

    b. decreases until a constant value of 0 is reached.

    c. doesn't change.

    d. increases indefinitely.

    e. increases until a constant value of 14 is reached.

WHY?

Chemists use the terms *acid* and *base* in a variety of ways, but General Chemistry focuses on the Brønsted-Lowry definition, which describes acids as proton donors and bases as proton acceptors. As explained in Activity 5-2, the terms *strong acid or base* and *weak acid or base* are used to express the extent to which these reactions proceed in water. You need to be able to determine equilibrium or final concentrations of hydronium ion—and therefore the pH—of various solutions involving both strong and weak acids and bases, because such reactions are involved in chemical, biological, and medicinal research and technology.

SUCCESS CRITERIA

- Recognize strong and weak acids and bases
- Calculate pH and pOH values for strong and weak acids and bases from concentrations and equilibrium constant expressions as necessary

PREREQUISITES

03-1 *Nomenclature: Naming Compounds*

05-2 *Introduction to Acid – Base Reactions*

15-1 *Dynamic Equilibrium and Le Châtelier's Principle*

15-2 *The Reaction Quotient & Equilibrium Constant*

16-1 *The pH Scale and Water Autoionization*

INFORMATION

Recall, and review if necessary, definitions of *acids* and *bases* presented in Activity 05-2. Recall as well that Activity 5-2 identified several common strong acids (proton donors) that you need to remember: HCl, HBr, HI, HNO_3, $HClO_4$ and H_2SO_4 (first proton only). There are other strong acids, but these six will be sufficient for now. Ionization reactions with these acids in aqueous solution are considered to go to completion; that is, to ionize completely. Most other acids are considered weak, meaning that they do not ionize completely in water, but instead establish an equilibrium involving the hydronium ion and the conjugate base. Weak acids, such as HF and H_3PO_4, are easy to recognize because the acidic hydrogen is listed first in their formulas. Organic acids, called carboxylic acids, can often be identified by the structural formula R-COOH, where R is the rest of the organic molecule, and the H at the end of the formula is the donated proton. A very common example of an organic acid is acetic acid, which found in vinegar and has the structural formula CH_3COOH. Its conjugate base is the acetate ion, CH_3COO^-. As you will see, weak acids vary in strength; the strength of an acid is quantitatively described by its *acid ionization equilibrium constant*, K_a.

MODEL 1	STRONG AND WEAK ACID pH CALCULATIONS

Strong Acid Ionization

$HCl(g) + H_2O(l) \rightarrow H_3O^+(aq) + Cl^-(aq)$ Ionization is complete: all of the HCl ionizes.

K_a is very large; $\Delta G°$ is a very large negative number.

$[H_3O^+]$ = moles HCl / volume of solution, so $[H_3O^+] = [HCl]$

$pH = -\log [H_3O^+]$. If $[HCl] = [H_3O^+] = 0.10$ M, pH = 1.0

Weak Acid Ionization

$CH_3COOH(aq) + H_2O(l) \rightarrow H_3O^+(aq) + CH_3COO^-(aq)$

Ionization is not complete: CH_3COOH is present in solution.

K_a is a small number, much less than 1; $\Delta G°$ is a positive number.

$$K_a = \frac{\left[H_3O^+\right]\left[CH_3COO^-\right]}{\left[CH_3COO\right]} = 1.76 \times 10^{-5}$$

In this example, the total initial concentration of CH_3COOH is 0.10 M.

R	CH_3COOH	H_2O	⇌	H_3O^+	CH_3COO^-
I	0.10 M	–		"0" M	0 M
C	– x			+ x	+ x
E	0.10 – x			x	x

$$K_a = 1.76 \times 10^{-5} = \frac{(x)(x)}{(0.10 - x)}$$

$$(1.76 \times 10^{-5})(0.10 - x) = x^2$$

$$1.76 \times 10^{-6} - (1.76 \times 10^{-5})x = x^2$$

$$0 = x^2 + (1.76 \times 10^{-5})x - 1.76 \times 10^{-6}$$

$$x = \frac{-(1.76 \times 10^{-5}) \pm \sqrt{(1.76 \times 10^{-5})^2 - (4)(1)(-1.76 \times 10^{-6})}}{2(1)}$$

$$x = 1.3 \times 10^{-3} \text{ and } -1.3 \times 10^{-3}$$

Only the positive value is reasonable for a concentration so:

$[H_3O^+] = 1.3 \times 10^{-3}$ M = $[CH_3COO^-]$ pH = $-\log(1.3 \times 10^{-3}) = 2.89$

$[CH_3COOH] = 0.10 - 1.3 \times 10^{-3}$, so its concentration is virtually unchanged.

1. What is meant by the statements in **Model 1** that ionization is "complete" or "not complete"?

2. According to **Model 1**, if a strong acid is added to water, what determines the pH of the resulting solution?

3. If a weak acid is added to water, according to **Model 1**, what determines the pH of the resulting solution?

4. What are five characteristics that distinguish a strong acid from a weak acid?

INFORMATION

In Model 1 it was assumed that the auto-ionization process of water contributed negligibly to the concentration of the hydronium ion, compared to the concentration produced by the ionization of the acid. If the values of K_a or the total concentration of the acid are very small, it may be necessary to include the initial concentration: 1.0×10^{-7} M.

In the calculations for hydronium ion concentration for the weak acid, it is often reasonable to simplify the mathematics greatly by assuming that the initial and equilibrium concentrations of the weak acid are the same, that is, that the value of x is small enough to be ignored when subtracted from the initial concentration. This is a good approximation only if the initial concentration of the weak acid is relatively large, on the order of 0.01 M or greater, and if the value of K_a is relatively small, on the order of 10^{-5} or smaller.

If those conditions are met, $x = \left[H_3O^+ \right] = \sqrt{Ka\left[HA \right]}$, where HA is a general weak acid.

If $\dfrac{x}{\left[HA \right]} \leq 0.05$ the approximation is usually acceptable, depending on the number of significant figures.

For the example in Model 1, $\left[H_3O^+ \right] = \sqrt{\left(1.76 \times 10^{-5} \right)\left(0.10 \right)} = 0.0013$ M which is equal to the result in **Model 1** within two significant figures.

EXERCISES

1. If 10.0 L of a 1.00 M solution of hydrochloric acid is diluted with water to a total volume of 1.00 L, what happens to the hydronium ion concentration and the pH?

2. Iodic acid, HIO_3, has a K_a value of 1.7×10^{-1}. Would it be acceptable to use the approximation method described in the Information section to calculate the pH of a 0.10 M solution of iodic acid? Why or why not?

GOT IT!

1. Which species has the highest concentration in an aqueous hydrocyanic acid solution: HCN, CN⁻, or H_3O^+? ($K_a = 6.2 \times 10^{-10}$)

MODEL 2 **STRONG AND WEAK BASE pH CALCULATIONS**

Strong Base Ionization

A strong base is typically a soluble metal hydroxide, like sodium hydroxide, that ionizes or dissociates in water to produce a metal cation and a hydroxide anion.

In water, $NaOH(s) \longrightarrow Na^+(aq) + OH^-(aq)$

Ionic dissociation is complete; all of the NaOH dissociates.

$\Delta G°$ is a large negative number.

$[OH^-]$ = moles NaOH/volume of solution

$pOH = -\log [OH^-]$, $pH = 14.00 - pOH$

Weak Base Ionization

A weak base, like ammonia, typically removes a proton from water to produce hydroxide ions.

$$NH_3(aq) + H_2O(l) \leftrightarrows NH_4^+(aq) + OH^-(aq)$$

Ionization is not complete; NH_3 is present in solution.

K_b is a small number, much less than 1; $\Delta G°$ is a large positive number.

$$K_b = \frac{\left[NH_4^+\right]\left[OH^-\right]}{\left[NH_3\right]} = 1.8 \times 10^{-5}$$

Hydroxide ion concentration can be approximated by

$$\left[OH^-\right] = \sqrt{K_b[B]}$$

where [B] is the initial concentration of the base. If $[NH_3] = 0.10$ M

$$\left[OH^-\right] = \sqrt{\left(1.8 \times 10^{-5}\right)\left(0.10\right)} = 0.0013 \text{ M}$$

$$pOH = 2.87, pH = 14.00 - 2.87 = 11.13$$

KEY QUESTIONS

5. What is meant by the statements in **Model 2** that ionization of a base is "complete" or "not complete"?

6. According to **Model 2**, if a strong base is added to water, what determines the pOH of the resulting solution?

7. If a weak base is added to water, according to **Model 2**, what determines the pOH of the resulting solution?

8. What are five characteristics that distinguish a strong base from a weak base?

EXERCISES

3. If a strong base is added to water, what happens to the hydroxide ion concentration and the pOH?

4. Calculate the pOH of a 0.50 M solution of the strong base NaOH.

5. Using the RICE table methodology developed in Activity 15-4, calculate the pOH of a 0.50 M ammonia solution ($K_b = 1.8 \times 10^{-5}$).

Got It!

2. Ammonium hydroxide (NH_4OH) is commonly advertised as an ingredient in some window cleaners. Order the following species from highest concentration to lowest in an aqueous solution of NH_4OH, and explain your reasoning. (Note that K_b for ammonia = 1.8×10^{-5}.)

$$NH_4OH \qquad NH_4^+ \qquad OH^- \qquad NH_3$$

3. Identify five similarities between acids and bases, with respect to their ionization reactions.

4. Identify five differences between acids and bases, with respect to their ionization reactions.

MODEL 3	K_a AND K_b FOR CONJUGATE ACID – BASE PAIRS ARE RELATED

Acid	pK_a	Conjugate Base	pK_b	pK_a + pK_b
HCN hydrocyanic acid	9.21	CN⁻ cyanide ion	4.79	14.00
$NH_4{}^+$ ammonium ion	9.25	NH_3 ammonia	4.75	14.00
H_2CO_3 carbonic acid	6.32	$HCO_3{}^-$ hydrogen carbonate ion	7.68	14.00
H_3O^+ hydronium ion	0.00	H_2O water	14.00	14.00
H_2O water	14.00	OH⁻ hydroxide ion	0.00	14.00

Note: $pK_a = - \log K_a$ and $K_a = 10^{-pK}$. The pK_a value is easier to write and is often easier to use than the K_a value, which involves scientific notation.

KEY QUESTIONS

9. What relationship between pK_a and pK_b can you identify in the table in **Model 3**?

10. Using your answer to Key Question 9, what relationship between K_a and K_b can you derive using the following property of logarithms: $\log(ab) = \log a + \log b$?

EXERCISES

9. Each member of your team should pick one acid-base pair from the table in **Model 3** and write the ionization reaction equations for the acid and the conjugate base. In addition, write the expressions for K_a and K_b in terms of the molar concentrations. Then combine the reaction equations by adding them, and multiply the K_a and K_b expressions together. Here is an example involving HCN and its conjugate base, CN^-:

(Equation 1) acid eq: $HCN + H_2O \leftrightarrows H_3O^+ + CN^-$ $K_a = \dfrac{[H_3O^+][CN^-]}{[HCN]} = 10^{-9.21}$

(Equation 2) base eq: $CN^- + H_2O \leftrightarrows HCN + OH^-$ $K_b = \dfrac{[HCN][OH^-]}{[CN^-]} = 10^{-4.79}$

(Equation 3) sum: $2\,H_2O \leftrightarrows H_3O^+ + OH^-$ $K_a \times K_b = [H_3O^+][OH^-] = 1.0 \times 10^{-14}$

Note: Equation (3) represents the autoionization of water, which has an equilibrium constant of 1.0×10^{-14} at 25 °C. In this auto or self-ionization reaction, water acts as both an acid (proton donor) and a base (proton acceptor).

GOT IT!

7. Each member of your team should get similar results for different acid-conjugate-base pairs in Exercise 9. Use these shared results to explain why $pK_a + pK_b = 14$ for any conjugate acid-base pair.

ACTIVITY 16-3 Finding pH for Salts and Polyprotic Acids and Bases

WHY?

Many interesting molecules act as either acids or bases (or both!) in chemical reactions. However, not all acids and bases have the same strength, as is clear from their various values of K_a or K. Larger values of K_a and K_b correspond to stronger acids and bases. The equations $K_a \times K_b = K_w$ and pH + pOH = pK_w indicate that the strengths of acids and their conjugate bases are inversely related; that is, the stronger the acid, the weaker its conjugate base. These ideas also apply to polyprotic acids and bases, which can often be treated as though they are monoprotic even though there are the same number of K_a or K_b values as the number of ionizable protons for the acid, or the number of protons that the base can accept. An understanding of the relative values of K_a and K_b are important in understanding titrations, the behavior of biological molecules, pharmaceuticals, and many other applications.

LEARNING OBJECTIVES

- Compare the relative strengths of acids and bases with their K_a, K_b, pK_a and pK_b values
- Compare the relative strengths of conjugate bases and acids that exist as salts
- Relate the extent of sequential ionization reactions in polyprotic acids and bases to corresponding values of K_a and K_b

SUCCESS CRITERIA

- Correctly calculate the pH values of salt solutions that contain the conjugates of weak acids or bases
- Correctly calculate the pH values of polyprotic acids and bases and their conjugates

PREREQUISITES

MODEL 1   REACTIONS AND pH CALCULATIONS OF CONJUGATE BASES AND ACIDS IN THE FORM OF SALTS

If 100.0 mL of 1.000 M acetic acid was combined with 100.0 mL of 1.000 M sodium hydroxide, the following quantitative neutralization reaction would take place.

$$HAc(aq) + NaOH(aq) \rightarrow H_2O(l) + NaAc(aq)$$

If all of the water produced was removed, the remaining solid, sodium acetate, could be collected. The pH of the product solution is above 7.00, because the acetate ion in the product is the conjugate base of acetic acid. In Activity 16-2, the value of the K_a of acetic acid was given as 1.76×10^{-5}.

Calculating the pH of the product solution, or an identical one made by dissolving the same amount of solid NaAc in the same amount of water, involves two steps: finding the concentration of the product, and then calculating the pH based on the equilibrium reaction.

Since the reaction is quantitative, there are no reactants left, and the concentration of the product is

$$[NaAc] = (100.0 \text{ mmol NaAc}) / (200.0 \text{ mL}) = 0.5000 \text{ M}$$

The pH of the solution depends on the reaction of the acetate ion as a base:

$$Ac^-(aq) + H_2O(l) \leftrightarrows HAc(aq) + OH^-(aq)$$

In order to calculate pH, the value of K_b for the acetate ion is needed.

$$K_b = K_w / K_a = (1.0 \times 10^{-14}) / (1.76 \times 10^{-5}) = 5.68 \times 10^{-10}$$

To find the pH, we first need to find $[OH^-]$. An equilibrium RICE table could be set up using 0.5000 M as the initial concentration of Ac. Assuming that the value of x is small, the result would be:

$$\left[OH^-\right] = \sqrt{K_b[B]} = \sqrt{\left(5.68 \times 10^{-10}\right)(0.5000)} = 1.68 \times 10^{-5} \text{ M}$$

which is much smaller than 0.5000 M, so the approximation is valid.

$$pOH = -\log(1.68 \times 10^{-5}) = 4.77 \text{ and pH} = 14.00 - 4.77 = \mathbf{9.23}$$

This is above 7.00, as predicted.

KEY QUESTIONS

1. How was the quantity of sodium acetate in the product solution calculated?

2. Why was a volume of 200.0 mL used when calculating the concentration of the product?

3. Why isn't the value of K_b simply equal to $1 / K_a$?

EXERCISE

1. The weak base ammonia, NH_3, has a pK_b of 4.75. What is the pH of a 0.250 M solution of ammonium chloride, NH_4Cl? Before you do any calculations, make a prediction about the pH to compare with your result.

Common oxo-acids like sulfuric acid, H_2SO_4, and phosphoric acid, H_3PO_4, or organic acids like oxalic acid, $H_2C_2O_4$, have more than one ionizable proton, which comes first in their formulae. Ionization reactions between these acids and water generally occur in several steps, since the K_a values for each of the protons differ, sometimes by several orders of magnitude. Phosphoric acid, for example, has pK_{a_1}, pK_{a_2}, and pK_{a_3} values of 2.2, 7.2, and 12.7 for the separate reactions:

$$H_3PO_4 + H_2O \leftrightarrows H_2PO_4^- + H_3O^+ \qquad pK_{a_1} = 2.15$$

$$H_2PO_4^- + H_2O \leftrightarrows HPO_4^{2-} + H_3O^+ \qquad pK_{a_2} = 7.20$$

$$HPO_4^{2-} + H_2O \leftrightarrows PO_4^{3-} + H_3O^+ \qquad pK_{a_3} = 12.35$$

When the pK_a values differ by 4 to 5 units, indicating that their respective K_a values differ by 4 to 5 orders of magnitude, the acid in its fully protonated form can be treated as a monoprotic acid using the calculation techniques already developed. It is interesting and important to note that when the solution pH = pK_a, the acid and conjugate base for the corresponding equilibrium have equal concentrations.

For most organic polyprotic acids, such as oxalic acid or phthalic acid, the separate pK_a values are usually closer in value, making the approximation in which the acid is treated as a monoprotic acid unreliable.

MODEL 2	POLYPROTIC ACIDS AND BASES AND THEIR CONJUGATES

Using the data from the Information section, the pH of a 0.10 M solution of phosphoric acid can be calculated using a RICE table method, making the approximation that $x \ll 0.10$. In this approximation, only the first acid dissociation is considered significant.

$$\left[H_3O^+\right] = \sqrt{K_{a_1}\left[H_3PO_4\right]}$$

$$K_{a_1} = 10^{-pKa_1} = 10^{-2.15} = 7.1 \times 10^{-3}$$

$$\left[H_3O^+\right] = \sqrt{\left(7.1 \times 10^{-3}\right)\left(0.10\right)} = 0.026 \text{ M}$$

However, (0.026) / (0.10) × 100% = 26%, so the approximation is not valid.

In the RICE table, "0" is used for the initial concentration of hydronium ion, since water's contribution of 1.0×10^{-7} M is expected to be negligible.

R	H_3PO_4	+	H_2O	$\leftrightarrows$	H_3O^+	+	$H_2PO_4^-$
I	0.10 M		–		"0" M		0 M
C	– x				+ x		+ x
E	0.10 – x				x		x

$$K_a = 7.1 \times 10^{-3} = \frac{(x)(x)}{(0.10 - x)}$$

$$(7.1 \times 10^{-3})(0.10 - x) = x^2$$

$$7.1 \times 10^{-4} - (7.1 \times 10^{-3})x = x^2$$

$$0 = x^2 + (7.1 \times 10^{-3})x - 7.1 \times 10^{-4}$$

$$x = \frac{-(7.1 \times 10^{-3}) \pm \sqrt{(7.1 \times 10^{-3})^2 - (4)(1)(-7.1 \times 10^{-4})}}{2(1)}$$

$$x = 2.3 \times 10^{-2} \quad \text{and} \quad -3.3 \times 10^{-2}$$

Only the positive value can describe a concentration so

$$[H_3O^+] = 2.3 \times 10^{-3} \text{ M} = [H_2PO_4^-] \quad \text{and} \quad pH = -\log(2.3 \times 10^{-2}) = \mathbf{1.64}$$

For the second equilibrium step, the additional concentration of hydronium ion is calculated with a second table, using 2.3×10^{-3} M as the initial concentration of both hydronium ion and dihydrogen phosphate ion.

R	$H_2PO_4^-$	+	H_2O	$\leftrightarrows$	H_3O^+	+	HPO_4^{2-}
I	2.3×10^{-3} M		–		2.3×10^{-3} M		0 M
C	$-x$				$+x$		$+x$
E	$2.3 \times 10^{-3} - x$				$2.3 \times 10^{-3} + x$		x

$$K_{a_2} = 10^{pK_{a_2}} = 10^{-7.20} = 6.3 \times 10^{-8} = \frac{(2.3 \times 10^{-3} + x)(x)}{(2.3 \times 10^{-3} - x)}$$

$$0 = x^2 + (2.300063 \times 10^{-3})x - 1.45 \times 10^{-10}$$

$$x = \frac{-(2.300063 \times 10^{-3}) \pm \sqrt{(2.300063 \times 10^{-3})^2 - (4)(1)(-1.45 \times 10^{-10})}}{2(1)}$$

$x = \mathbf{6.3 \times 10^{-8}}$ Only the positive value can describe a concentration. Our result indicates that the increase in hydronium ion from the second step is negligible.

KEY QUESTIONS

4. Why was it not appropriate to use the simpler square root equation to find $[H_3O^+]$ for the first ionization reaction in **Model 2**, while it was acceptable for the second ionization reaction?

5. How would the calculations for the pH of a 0.10 M solution of H_2SO_4 differ from the calculations in the model?

EXERCISE

2. Calculate the pH of a 0.250 M solution of sodium phosphate, Na_3PO_4.

PROBLEM

1. A general chemistry textbook problem about solubility rules states that a solution of ammonium phosphate is mixed with another solution containing sodium carbonate. The question asks whether there will be a precipitation reaction when the solutions are combined. What other chemical reactions might be involved in the original solutions and/or the mixture? Explain.

WHY?

Solutions of a weak acid and its conjugate base are called *buffers*. Buffer solutions have specific pH values, which resist change due to dilution or the addition of strong acids or bases. This buffering action is a vital part of many biological and environmental systems. For example, blood cells function only within a narrow pH range, and the pH of the world's oceans is decreasing due to rising atmospheric CO_2 levels that increase the oceanic concentration of carbonic acid, H_2CO_3. Carbonic acid is the acid component of the oceans' buffer system. When working with pH-sensitive reactions, you need to introduce an appropriate buffer to control the pH.

LEARNING OBJECTIVES

- Understand the nature and properties of buffer solutions
- Recognize the origin and use of the Henderson-Hasselbalch equation
- Understand the key reactions between strong acids or bases and buffers

SUCCESS CRITERIA

- Accurately calculate the pH of buffer solutions
- Determine the effects of the addition of a strong acid or base on the pH of a buffer solution

PREREQUISITES

04-4   *Solving Solution Stoichiometry Problems*

16-2   *Strong and Weak Acids and Bases*

16-3   *Finding pH for Salts and Polyprotic Acids and Bases*

INFORMATION

When calculating the pH of buffer solutions, it is necessary to use the appropriate K_a value for the weak acid–conjugate base pair. In a buffer system, the concentrations of the acid and base species are similar or equal. In most cases, the change in concentration (the x value in the corresponding RICE table) is very small, and can be neglected as the buffer reaches equilibrium. Consequently, the initial concentrations are good approximations of their final concentrations.

| MODEL 1 | BUFFER pH CALCULATIONS AND THE HENDERSON-HASSELBALCH EQUATION |

The pH of an acetic acid–sodium acetate buffer: [HAc] = 0.20 M   [NaAc] = 0.15 M

$$K_a = [H_3O^+][Ac^-] / [HAc] = 1.76 \times 10^{-5}$$

$$= (x)(0.15 + x) / (0.20 - x) \qquad \textit{if } x << 0.15 \textit{ M and } 0.20 \textit{ M}$$

$$[H_3O^+] = x$$

$$= (1.76 \times 10^{-5})(0.20)/(0.15)$$

$$= 2.35 \times 10^{-5} \textit{ M} \qquad \textit{<< 0.15 M and 0.20 M,}$$
$$\textit{so the approximation is valid.}$$

$$pH = - \log (2.35 \times 10^{-5}) = \mathbf{4.63}$$

$K_a = [H_3O^+][Ac^-]/[HAc] = 1.76 \times 10^{-5}$ can be rearranged to: $[H_3O^+] = K_a [HAc]/[Ac^-]$

Taking the –log of both sides gives: $pH = pK_a + \log([Ac^-] / [HAc])$

or more generally: $\mathbf{pH = pK_a + \log([A^-] / [HA])}$

This is known as the Henderson-Hasselbalch equation, where HA is a general weak acid and A⁻ is its conjugate base.

Applying this equation to the same buffer system, and assuming x << 0.15 M and 0.20 M:

$$pK_a = - \log (1.76 \times 10^{-5}) = 4.75$$
$$pH = 4.75 + \log[(0.15) / 0.20)] = \mathbf{4.63} \qquad \textit{which is the \textbf{same pH} given by}$$
$$\textit{the } K_a \textit{ equilibrium equation.}$$

KEY QUESTIONS

1. According to both the K_a equilibrium equation and the Henderson-Hasselbalch equation, under what conditions (what concentrations of the weak acid and conjugate base) does the pH of the buffer equal the pK_a of the weak acid?

2. Write an equation similar to the Henderson-Hasselbalch equation that can be used to determine the pOH of a buffer that is made from a weak base, B, and its conjugate acid, BH⁺.

3. According to the equations that can be used to calculate the pH of a buffer, how do the concentrations of the conjugate acid and base affect the overall pH?

4. In the laboratory you are provided with a stock pH 7.00 buffer solution. What will happen to the pH if the buffer is diluted by a factor of 10?

INFORMATION

Buffers resist changes in pH due to the addition of strong acids or strong bases, but only to a certain extent. If some amount of a strong acid is added to a buffer that contains a greater number of moles of its weak base, the strong acid reacts with a portion of this conjugate base. As it does, the strong acid, as the limiting reactant, is completely consumed; the concentration of the conjugate base decreases, and the concentration of the weak acid is increased. The resulting pH change depends upon the resulting ratio of ($[A^-]$ / $[HA]$). If more moles of strong acid are added to the buffer than there are moles of the conjugate base, the capacity of the buffer to resist pH change is exceeded. The excess moles of strong acid determine the pH of the resulting solution. A similar situation exists if a strong base is added to the buffer, since it reacts quantitatively with the weak acid component of the buffer.

Buffers are most effective when the desired pH is very close to the pK_a of the weak acid, and the concentrations of both the weak acid and conjugate base are relatively large; for instance 0.10 M to 1.0 M.

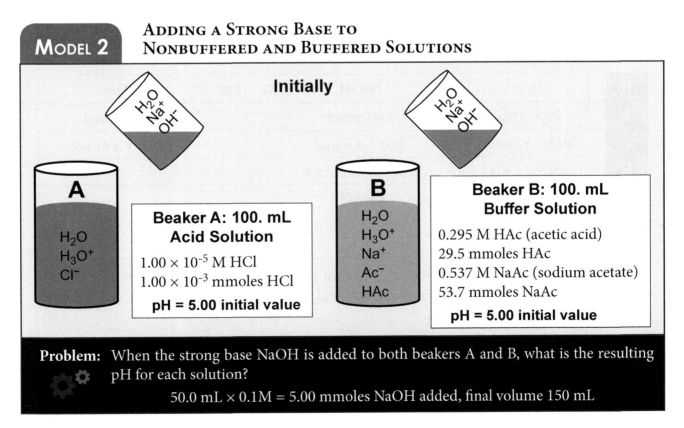

MODEL 2 **ADDING A STRONG BASE TO NONBUFFERED AND BUFFERED SOLUTIONS**

Initially

H₂O + Na⁺ OH⁻

H₂O + Na⁺ OH⁻

A

H₂O
H₃O⁺
Cl⁻

Beaker A: 100. mL Acid Solution

1.00×10^{-5} M HCl
1.00×10^{-3} mmoles HCl

pH = 5.00 initial value

B

H₂O
H₃O⁺
Na⁺
Ac⁻
HAc

Beaker B: 100. mL Buffer Solution

0.295 M HAc (acetic acid)
29.5 mmoles HAc
0.537 M NaAc (sodium acetate)
53.7 mmoles NaAc

pH = 5.00 initial value

Problem: When the strong base NaOH is added to both beakers A and B, what is the resulting pH for each solution?

50.0 mL × 0.1M = 5.00 mmoles NaOH added, final volume 150 mL

Beaker A: Strong Acid + Strong Base: an Unbuffered Solution

A reaction table to find the limiting reactant and final concentrations of products is used. Remember that these tables are in terms of moles or mmoles, and assume that the reaction goes to completion.

R	HCl	+	NaOH	$\leftrightarrows$ H$_2$O	+	NaCl
I	1.00×10^{-3} mmol		5.00 mmol	–		0 mmol
C	$1.00 \times 10^{-3} - x$ mmol		$5.00 - x$ mmol	–		$+ x$ mmol
E	$1.00 \times 10^{-3} - x = 0$		$5.00 - x$ mmol $= 0$			
LR	$x = 1.00 \times 10^{-3}$		$x = 5.00$			
Left	1.00×10^{-3} $- 1.00 \times 10^{-3} = 0$		$5.00 - 1.00 \times 10^{-3}$ $= 5.00$ mmol			1.00×10^{-3} mmol, but irrelevant

Since the x value for HCl is clearly much smaller than that of NaOH, HCl is the limiting reactant, and is completely consumed. The amount of NaOH = 5.00 – 0.001 mmoles, so the amount of NaOH is virtually unchanged.

The pH will be determined by the total concentration of NaOH in the total volume of solution.

$$[OH^-] = 5.00 \text{ mmoles} / 150 \text{ mL} = 0.0333 \text{ M}, \text{pOH} = 1.48$$

$$\text{final pH} = 14.00 - 1.48 = \textbf{12.52}$$

Beaker B: Buffer + Strong Base: a Buffered Solution

It is first necessary to determine which buffer component reacts with the added strong base. Strong bases react completely with weak acids, so the extent of this reaction will determine the new concentrations of both the weak acid and its conjugate base.

A reaction table to find the limiting reactant and final concentrations of products is used. Remember that these tables are in terms of moles or mmoles, and assume that the reaction goes to completion.

R	HAc	+	NaOH	$\leftrightarrows$ H$_2$O	+	NaAc
I	29.5 mmol		5.00 mmol	–		53.7 mmol
C	$29.5 - x$ mmol		$5.00 - x$ mmol	–		$53.7 + x$ mmol
E	$29.5 - x = 0$		$5.00 - x$ mmol $= 0$			
LR	$x = 29.5$		$x = 5.00$			
Left	$29.5 - 5.00$ $= 24.5$ mmol		$5.00 - 5.00 = 0$			$53.7 + 5.00$ $= 58.7$ mmol

Since the x value for NaOH is smaller than that of HAc, NaOH is the limiting reactant, and is completely consumed. There are 24.5 mmol of HAc and 58.7 mmol of Ac$^-$ remaining. Concentrations can be found by dividing by 150 mL.

$$[Ac^-] = 0.391 \text{ M}, [HAc] = 0.163 \text{ M}.$$

The pH of the new buffer can now be calculated:

$$\text{final pH} = 4.75 + \log[(0.391) / (0.163)] = \textbf{5.13}$$

5. What are the initial pH values for the solutions in beaker A and beaker B?

6. When 5.00 mmoles of hydroxide are added to beaker A, what will be the pH of the resulting solution?

7. When 5.00 mmoles of hydroxide are added to beaker B, what will be the pH of the resulting solution?

8. When 5.00 mmoles of hydroxide are added to beaker B, why does the equilibrium amount of hydroxide in the resulting solution remain small, and why does the pH change only slightly?

9. What equilibrium constant equation is used to calculate the hydronium ion concentration for the

10. What parameters are most important for determining the initial and the final hydronium ion concentration of solution B?

11. Suppose you wanted to make a third buffer solution (solution C) with pH = 3. How would you do so?

EXERCISES

1. Identify the essential components of a buffer solution.

2. Calculate the pH of a solution 0.250 M acetic acid and 0.330 M sodium acetate.

3. Calculate the ratio of $[NH_3]/[NH_4^+]$ needed in an ammonia/ammonium chloride buffer to produce a pH of 9.25.

4. Identify which of the following result in buffer solutions when equal volumes of the two solutions are mixed.

   a. 0.1 M KNO_3 and 0.1 M HNO_3

   b. 0.1 M HCl and 0.1 M NH_3

   c. 0.2 M HCl and 0.1 M NH_3

d. 0.1 M HCl and 0.2 M NH$_3$

PROBLEM

1. A buffer for maintaining a pH of 2.00 is needed for an experiment. A possible solution for this purpose is the phosphoric acid–sodium dihydrogen phosphate buffer system based on the equilibrium:

$$H_3PO_4 + H_2O \rightleftarrows H_2PO_4^- + H_3O^+ \qquad pK_{a1} = 2.15$$

Is it appropriate to use the Henderson-Hasselbalch equation to calculate the relative amounts of phosphoric acid and sodium dihydrogen phosphate needed to make 1.00 L of this buffer? Assume the initial concentration of the phosphoric acid is 0.100 M and the necessary concentration of sodium dihydrogen phosphate must be calculated. Explain your reasoning.

WHY?

As described in Activity 05-4, titrations involve gradual addition of titrant solutions to analyte solutions until an endpoint volume is determined, which is then used to determine the concentration of the analyte. While colored chemical indicators are often used to determine the exact amount of titrant required in an acid-base titration, a pH meter can be used instead. To do this, however, you must use your knowledge of acid-base reactions and equilibria to predict the pH of the analyte solution at various points in the titration and to predict the endpoint pH. Titrations are widely used for quantitative chemical analyses in industrial quality control, clinical and environmental laboratories, and chemical research.

LEARNING OBJECTIVES

- Understand why titration curves have particular shapes
- Understand the information presented by a titration curve

SUCCESS CRITERIA

- Accurately calculate of the pH at various points during a titration
- Correctly identify major species in solution during a titration

PREREQUISITES

05-4 *Analytical Titrations*

16-2 *Strong and Weak Acids and Bases*

16-3 *Finding pH for Salts and Polyprotic Acids and Bases*

17-1 *Buffer Solutions*

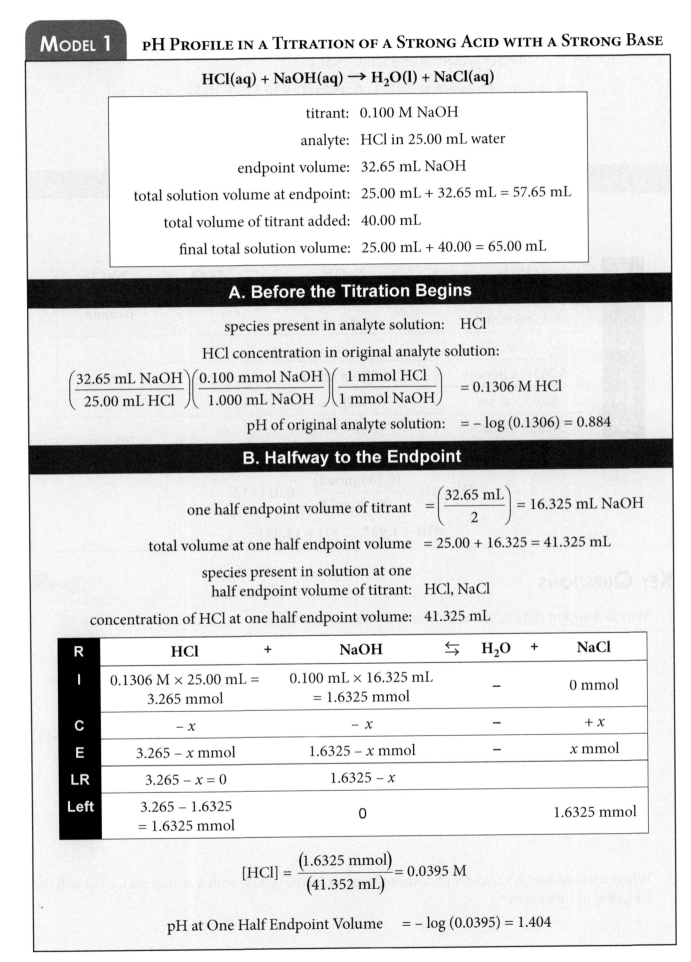

MODEL 1 — pH PROFILE IN A TITRATION OF A STRONG ACID WITH A STRONG BASE

$$HCl(aq) + NaOH(aq) \rightarrow H_2O(l) + NaCl(aq)$$

titrant: 0.100 M NaOH

analyte: HCl in 25.00 mL water

endpoint volume: 32.65 mL NaOH

total solution volume at endpoint: 25.00 mL + 32.65 mL = 57.65 mL

total volume of titrant added: 40.00 mL

final total solution volume: 25.00 mL + 40.00 = 65.00 mL

A. Before the Titration Begins

species present in analyte solution: HCl

HCl concentration in original analyte solution:

$$\left(\frac{32.65 \text{ mL NaOH}}{25.00 \text{ mL HCl}}\right)\left(\frac{0.100 \text{ mmol NaOH}}{1.000 \text{ mL NaOH}}\right)\left(\frac{1 \text{ mmol HCl}}{1 \text{ mmol NaOH}}\right) = 0.1306 \text{ M HCl}$$

pH of original analyte solution: $= -\log(0.1306) = 0.884$

B. Halfway to the Endpoint

one half endpoint volume of titrant $= \left(\dfrac{32.65 \text{ mL}}{2}\right) = 16.325 \text{ mL NaOH}$

total volume at one half endpoint volume $= 25.00 + 16.325 = 41.325 \text{ mL}$

species present in solution at one half endpoint volume of titrant: HCl, NaCl

concentration of HCl at one half endpoint volume: 41.325 mL

R	HCl	+	NaOH	⇄ H₂O	+	NaCl
I	0.1306 M × 25.00 mL = 3.265 mmol		0.100 mL × 16.325 mL = 1.6325 mmol	−		0 mmol
C	− x		− x	−		+ x
E	3.265 − x mmol		1.6325 − x mmol	−		x mmol
LR	3.265 − x = 0		1.6325 − x			
Left	3.265 − 1.6325 = 1.6325 mmol		0			1.6325 mmol

$$[HCl] = \frac{(1.6325 \text{ mmol})}{(41.352 \text{ mL})} = 0.0395 \text{ M}$$

pH at One Half Endpoint Volume $= -\log(0.0395) = 1.404$

C. At the Endpoint

species present at endpoint: NaCl

total volume of solution at endpoint: $= 25.00 + 16.325 + 16.325 = 57.65$ mL

concentration at endpoint: 3.265 mmol / 57.65 mL = 0.0566 M

pH at endpoint: 7.00

D. At the Final Volume

species present in final solution: NaCl, NaOH

concentration of NaOH in final solution:

R	HCl	+	NaOH	$\leftrightarrows$	H$_2$O	+	NaCl
I	0.1306 M × 25.00 mL = 3.265 mmol		0.100 mL × 40.00 mL = 4.00 mmol		–		0 mmol
C	– x		– x		–		+ x
E	3.265 – x mmol		4.00 – x mmol		–		x mmol
LR	3.265 – x = 0		4.00 – x				
Left	3.265 – 1.6325 = 1.6325 mmol		4.00 – 3.265 = 0.735 mmol				3.265 mmol

$$[\text{NaOH}] = \frac{(0.735 \text{ mmol})}{(65.00 \text{ mL})} = 0.0113 \text{ M}$$

$$\text{pOH} = 1.947 \qquad \text{pH} = 12.053$$

KEY QUESTIONS

1. Why is it appropriate to use a RICE table for these calculations?

2. The concentration of NaCl was never taken into account in any of the calculations of pH in **Model 1**. Why not?

3. When a strong acid is titrated with a strong base, or a strong base with a strong acid, what will the endpoint pH always be?

4. Why is it necessary to use the total solution volume in the calculations of pH?

EXERCISE

1. Based on the approach used in **Model 1**, complete Table 1 by calculating the missing pH values based on the major species in solution. Plot those points on the graph and sketch the corresponding pH curve for this titration.

Table 1   **Calculated pH Values for Major Species in Solution (Model 1)**

Added Volume of Base	Major Species in Solution	Major Species, mmol	Total Volume, mL	Concentration of Major Species, M	pH
0 mL	HCl	3.265	25.00	0.1306	0.884
10 mL					
16.33	HCl NaCl	1.6325 1.6325	41.33	0.0395 0.0395	1.403
32.00					
32.65	NaCl	3.265	57.65	0.0566	7.000
33.00					
40.00	NaOH NaCl	0.735 3.265	65.00	0.0113 0.0502	12.053

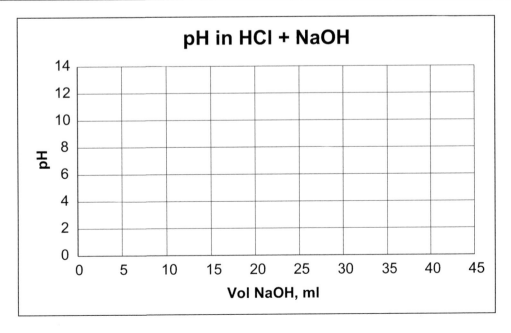

pH in HCl + NaOH

INFORMATION

In **Model 1**, the product NaCl did not affect pH at any point, because it is a salt made up of conjugates of strong acids and bases, and is therefore neither acidic nor basic. The pH at the end point was neutral (equal to 7.000), the same as pure water. If the titration involved adding a strong acid to a strong base, the endpoint would again have a neutral pH, but the curve would be inverted, starting at high pH and proceeding to low pH as acid was added and eventually became the excess reagent.

For the titration of a weak acid with a strong base or the titration of a weak base with a strong acid, the pH of the analyte must take into account the K_a or K_b as well as the concentration. Also, the product salt is NOT neutral, since it is the conjugate base or acid of the reactant. Since a complete reaction is necessary for titration, only strong acids and bases are added to analyte, regardless of whether the analyte itself is strong or weak. Titrations of weak acids with weak bases are not effective. As the titration progresses and the conjugate of the analyte is generated as a product, a buffer solution is formed. The endpoint pH, therefore, depends on the concentration of the remaining analyte and the product. At the endpoint, the pH is determined by the concentration of the conjugate base or acid, so it is NOT equal to 7.00.

MODEL 2	pH PROFILE OF THE TITRATION OF A WEAK BASE WITH A STRONG ACID

Analyze the pH values resulting from the titration of a weak base (ammonia, NH_3) with a strong acid (hydrochloric acid, HCl) by completing Exercises 1 and 2, and Table 2. In this case, a reaction table like the one used in **Model 1** can be used to find the concentrations of any leftover reactants and any products of the reaction. Alternate methods will be required to calculate the pH during the reaction, before reaching the endpoint.

Assume that you are titrating 100.0 mL of 0.0500 M ammonia ($K_b = 1.8 \times 10^{-5}$) with 0.100 M hydrochloric acid.

Construct a graph of your results from the data you enter in Table 2.

EXERCISES

2. Before any HCl has been added, the pH of the solution is determined by the reaction of ammonia with water. Write the reaction of ammonia with water and the corresponding equilibrium constant expression. Determine the value for K_b (the equilibrium constant).

3. Write the reaction equation for the reaction of NH_4^+ with water, and the corresponding equilibrium constant expression. This equilibrium constant is called K_a, the acid ionization constant. Remember that $K_a = K_w / K_b$, and also determine the value for K_a.

Calculations Needed to Complete Table 2

Calculation 1 **Before adding HCl:**

Calculation 2 **After the addition of 10 ml of HCl:**

Calculation 3 **After the addition of 25 ml of HCl:**

Chapter 17: Buffers, Titrations, and Solubility Equilibria

Calculation 4   After 49.5 mL of HCl has been added:

Calculation 5   After 50 ml of HCl has been added:

Calculation 6   After 50.5 mL of HCl has been added:

Calculation 7   After 60 ml of HCl has been added:

Calculation 8   After 75 ml of HCl has been added:

Table 2   **Calculated Data for the Titration of 100.0 mL 0.0500 M Ammonia with 0.10 M Hydrochloric Acid**

Situation	Major Species	$[H_3O^+]$	pH
before adding HCl			
10.0 mL HCl added			
25.0 mL HCl added			
49.5 mL HCl added			
50.0 mL HCl added			
50.5 mL HCl added			
60.0 mL HCl added			
75.0 mL HCl added			

Graph of the Titration of
100.0 mL 0.0500 M Ammonia with 0.10 M Hydrochloric Acid

Plot pH on the *y*-axis and volume of HCl added on the *x*-axis. (Remember that a high-quality graph has a title, labeled axes, and a scale, and that points are plotted with small circles around them to make them easier to see, with a smooth line is drawn through the points.)

KEY QUESTIONS

5. What is the net ionic chemical reaction equation for the reaction of hydrochloric acid with ammonia?

6. Why is the halfway point for this titration considered to be the point at which 25.0 mL of HCl have been added?

7. What are the relative concentrations of NH_4^+ and NH_3 at the halfway point? Provide a qualitative answer, not numbers.

8. Why is the stoichiometric point for this titration considered to be the point at which 50.0 mL of HCl have been added?

9. At the stoichiometric point, what are the relative concentrations of NH_4^+ and NH_3? Provide a qualitative answer, not numbers.

10. At what point in the titration is the pH equal to the pK_a of the weak acid involved in the titration? Why?

11. What are the major species present in the solution after the stoichiometric point has been reached, e.g., at 75 mL?

–

12. Why does the pH change so abruptly near the stoichiometric point?

Got It!

1. For each of the following types of acid-base titrations, roughly sketch the dependence of pH on the volume of titrant solution added. Also describe in words how and why the pH changes as it does. The analyte is listed first, the titrant second.

   a. strong acid + strong base

b. weak acid + strong base

c. strong base + strong acid

Chapter 17: Buffers, Titrations, and Solubility Equilibria

d. weak base + strong acid

Solubility and the Solubility Product

WHY?

Many ionic compounds are very soluble in water, while others are very insoluble. For example, sodium chloride, NaCl, and silver chloride, AgCl, appear to be very similar compounds. Yet sodium chloride is very soluble, and silver chloride is very insoluble in water. This is a result of the balance that is struck between the lattice energy of the ions in the solid, the hydration energy when the ions go into solution, and the change in entropy when the solution is formed. An equilibrium constant, which is called the *solubility product constant* (K_{sp}), is used to indicate the extent to which any ionic solid is soluble. The solubility product constant enables one to predict how much of any compound will dissolve, how conditions like pH and the presence of other ions affect solubility, and, when given several ionic compounds, to predict which will precipitate first.

LEARNING OBJECTIVES

• Learn how to write the expression for the solubility product constant

• Use the solubility product constant in solving quantitative problems involving solubility

SUCCESS CRITERIA

• Correctly predict solubilities given solubility product constants

• Calculate accurate values for solubility product constants given quantitative data on solubilities

PREREQUISITES

05-1   *Dissociation and Precipitation Reactions*

15-2   *The Reaction Quotient & Equilibrium Constant*

15-3   *Free Energy and Chemical Equilibrium*

15-4   *Solving Equilibrium Problems: The RICE Table Methodology*

INFORMATION

This activity neglects several complicating factors in order to make the connection between solubility and the solubility product constant as simple as possible. In reality, when ionic solids dissociate in water, the resulting ions can react with water. For example, when lead(II) sulfate ($PbSO_4$) dissociates, the sulfate ion (SO_4^{2-}) reacts with water to produce hydrogen sulfate (HSO_4^-). Ion pairs can also be formed. For example, when lead(II) chloride dissociates in water, the resulting solution contains $PbCl^+$ as well as Pb^{2+} and Cl^-. Such effects, which are ignored in this activity, make the compounds more soluble than solubility product constants would predict.

Four salts (silver chloride, sodium chloride, lead(II) iodide, and lead(II) sulfate, are added to water in four separate beakers until the solutions are saturated and excess solid remains in each beaker.

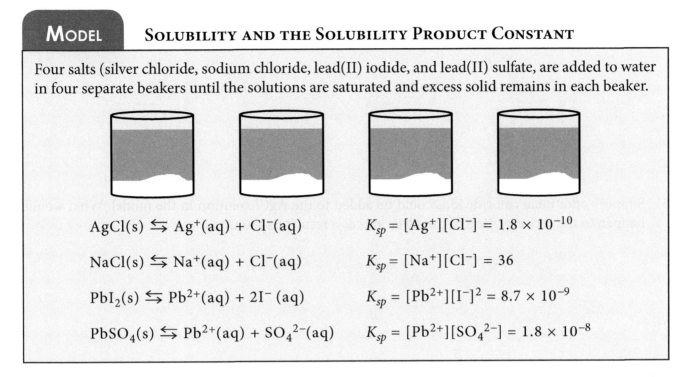

$$AgCl(s) \leftrightharpoons Ag^+(aq) + Cl^-(aq) \qquad K_{sp} = [Ag^+][Cl^-] = 1.8 \times 10^{-10}$$

$$NaCl(s) \leftrightharpoons Na^+(aq) + Cl^-(aq) \qquad K_{sp} = [Na^+][Cl^-] = 36$$

$$PbI_2(s) \leftrightharpoons Pb^{2+}(aq) + 2I^-(aq) \qquad K_{sp} = [Pb^{2+}][I^-]^2 = 8.7 \times 10^{-9}$$

$$PbSO_4(s) \leftrightharpoons Pb^{2+}(aq) + SO_4{}^{2-}(aq) \qquad K_{sp} = [Pb^{2+}][SO_4{}^{2-}] = 1.8 \times 10^{-8}$$

KEY QUESTIONS

1. How is the form of the solubility product constant similar to, and how does it differ from, the standard form of an equilibrium constant for a chemical reaction equation?

2. Is silver chloride or sodium chloride more soluble in water? Explain how specific information given in the model supports your conclusion.

3. In general, when comparing solubility product constants to predict solubilities as you did in Key Question 2, what information needs to be taken into account, in addition to the K_{sp} values? Carefully consider the situation for lead(II) iodide and lead(II) sulfate in the model.

4. Suppose chloride ions could be removed selectively from the AgCl solution in the model. What would happen to the concentration of the silver ion as a result? Explain.

5. Suppose additional chloride ions could be added to the AgCl solution in the model. What would happen to the concentration of the silver ions as a result? Explain.

EXERCISES

1. What is the solubility of barium sulfate ($K_{sp} = 1.1 \times 10^{-10}$) in moles/L and in g/L?

2. A saturated solution of MgF_2 contains 1.2×10^{-3} M Mg^{2+}. What is the value of its solubility product constant?

3. How much lead(II) chloride will dissolve in a solution of 0.25 M sodium chloride? The K_{sp} for lead (II) chloride is 1.7×10^{-5}.

Problems

1. In 2014 in Flint, Michigan, officials decided to switch water sources, which began the city's problems with the water supply. Failing to treat the water made the lead pipes start leaching lead into the water supply. Lead is measured in parts per billion (ppb) and researchers have found lead in homes in Flint at levels that the EPA qualifies as toxic waste.

   If the lead came from lead(II) hydroxide ($K_{sp} = 2.8 \times 10^{-16}$), opposed to other lead compounds that could also be found dissolved in water, what would be the pH levels of the water in Flint be, if they found 5,000 ppb of lead?

2. In order to prevent the lead pipes from leaching into the water supply, most cities keep the water at a basic pH level. In Flint, when they switched water sources, they stopped treating the water. (Maintaining a relatively high pH level of water from treatment plants—around a pH of 10—keeps a protective layer inside the pipes from dissolving).

   Based on the equilibrium constant expression for lead (II) hydroxide, how much lower is the maximum possible lead(II) ion concentration at pH 10.0 than at pH 5.0? To answer this question, calculate the ratio: (concentration at pH 10.0) / (concentration at pH 5.0).

3. In a mining operation it necessary to separate copper (I) ions from lead (II) ions. The concentration of each of these ions in solution is 1.0×10^{-3} M. The strategy is to gradually add sodium iodide to precipitate one iodide salt before the other.

   a. Which salt precipitates first, CuI ($K_{sp} = 5.1 \times 10^{-12}$) or PbI$_2$ ($K_{sp} = 8.7 \times 10^{-9}$)?

   b. What is the iodide concentration necessary to begin the precipitation of each salt?

c. What is the remaining concentration of the ion that began precipitating first at the point where the other ion begins to precipitate?

d. What percent of the original copper concentration remains in solution?

4. Although silver chloride is very insoluble in water, it can be dissolved in a solution of aqueous ammonia due to the following series of reactions leading to the formation of the diamminesilver complex ion.

$$AgCl(s) \leftrightarrows Ag^+(aq) + Cl^-(aq) \qquad\qquad K_{sp} = 1.8 \times 10^{-10}$$

$$Ag^+(aq) + NH_3(aq) \leftrightarrows Ag(NH_3)^+ (aq) \qquad\qquad K_{f1} = 2.1 \times 10^3$$

$$Ag(NH_3)^+ (aq) + NH_3(aq) \leftrightarrows Ag(NH_3)_2^+ (aq) \qquad K_{f2} = 8.2 \times 10^3$$

a. Determine the value of the equilibrium constant for the overall reaction:

$$AgCl(s) + 2\,NH_3(aq) \leftrightarrows Ag(NH_3)_2^+(aq) + 2Cl^-(aq)$$

b. Determine the amount (moles) of silver chloride that will dissolve in 1.0 L of a 10.0 M ammonia solution.

ACTIVITY 18-1

Standard Reduction Potentials and Electrochemical Cells

WHY?

Electron-transfer reactions, also known as *redox reactions*, were described in Activity 5-3. These reactions have wide-ranging importance in biochemical and environmental systems. The half-reaction concept used in balancing redox reactions, which involves electrons as products of an oxidation reaction and as reactants in a reduction reaction, is essential to the description of *electrochemical cells*. In these cells, the half-reactions are kept separate from each other, and require that the electrons that are transferred do so by flowing through an external circuit, creating direct current electricity. Common battery designs are based on these principles. The potential energy states of the electrons in the reactants and products are measured and tabulated as *standard reaction potentials*, in volts, in comparison to the *standard hydrogen electrode.*

LEARNING OBJECTIVES

- Understand the measurement process and meaning of the term standard reduction potential
- Understand how an electrochemical cell produces electricity by a spontaneous redox reaction
- Understand that Standard Reduction Potentials are measured in comparison to the Standard Hydrogen Electrode

SUCCESS CRITERIA

- Identify the components of a cell and sites of oxidation and reduction
- Distinguish between the terms *standard reduction potential* and *standard cell potential*
- Calculate standard cell potentials based on standard reduction potentials
- Determine the reactants, products, and spontaneity of a redox reaction based on standard reduction potentials
- Calculate change in free energy based on standard cell potentials

PREREQUISITES

05-3 *Electron Transfer Reactions*

15-3 *Free Energy and Chemical Equilibrium*

INFORMATION

A redox reaction can be separated into two half-reactions; an oxidation reaction and a reduction reaction. In *electrochemical cells*, each reaction occurs in a different compartment, and electrons are transferred through a wire from one compartment to the other. The term *anode* is used to indicate the solid, often a metal, where the oxidation reaction takes place. Since electrons are being lost at this *electrode,* as they move into the wire, a positive charge develops so that anions are attracted to the anode. The term *cathode* is used to indicate the solid, also often a metal, where the reduction reaction occurs. Since electrons are being transferred through the wire, a negative charge develops, and cations are attracted to the cathode. It is necessary to maintain a charge balance by allowing ions to flow between the compartments; very often, this is done with a *salt bridge* that contains an inert salt that is neither oxidized nor reduced in the cell.

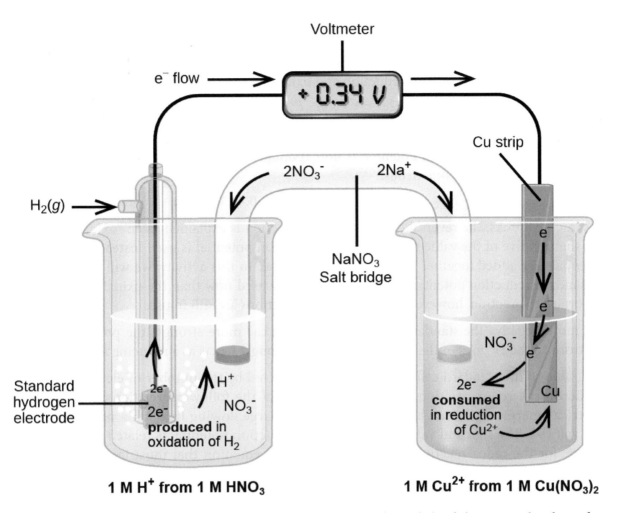

The *standard reduction potential*, $E°_{red}$, is a measurement (in volts) of the strength of a reduction reaction's ability to attract electrons from hydrogen gas in the *standard hydrogen electrode (SHE)*. As with enthalpy and free energy, it is necessary to establish a reference point for these values. The enthalpy and free energy of formation of an element are defined as zero. Likewise, the SHE is used as the reference point for electrochemical reactions. It is composed of a platinum metal anode at which hydrogen gas is oxidized to hydronium ions under standard conditions (pressure = 1 atm, temperature = 25 °C, concentrations = 1 M). As the standard reference value, the $E°_{red}$ of the SHE is defined as 0 V exactly.

The reaction at the cathode also takes place under standard conditions in a measurement of standard reduction potential. An example would be the two-electron reduction of the copper(II) ion at 1 M concentration to form solid copper metal, written

$$Cu^{2+}(aq) + 2e^- \longrightarrow Cu(s) \quad E°_{red} = +0.337 \text{ V (cathode)}$$
$$H_2(g) \longrightarrow 2H^+(aq) + 2e^- \quad E°_{red} = 0.00 \text{ V (anode)}$$

This is a spontaneous process with the SHE acting as the anode, so its reaction equation is reversed. The positive value of the measured $E°_{red}$ indicates the spontaneity of the overall redox reaction. The more positive the value of the $E°_{red}$, the more strongly the oxidized species on the reactants side attracts electrons (and, therefore, the stronger it acts as an oxidizing agent). In general, the standard cell potential, $E°_{cell}$, is defined as the $E°_{red}$ of the cathode minus the $E°_{red}$ of the anode, or

$$E°_{cell} = E°_{cat} - E°_{an} = +0.337 \text{ V} - 0.00 \text{ V} = +0.337 \text{ V}$$

For negative values of $E°_{red}$, the flow of electrons is reversed, since the product of the reduction reaction is able to transfer electrons to H^+ in the SHE. For the spontaneous version of the process, then, the SHE would become the *cathode*, since electrons are flowing to it spontaneously. The actual anode would be the half-reaction being measured, since oxidation would take place there. An example of this would be the reduction reaction of iron(II) ion to the solid metal iron. To be consistent, however, the written reduction and oxidation reactions are used as before.

$$Fe^{2+}(aq) + 2e^- \rightarrow Fe(s) \quad E^\circ_{red} = -0.44 \text{ V (cathode, but acts as anode)}$$

$$H_2(g) \rightarrow 2H^+(aq) + 2e^- \quad E^\circ_{red} = 0.00 \text{ V (anode, but acts as cathode)}$$

In this case, the standard cell potential is non-spontaneous, since the measured E°_{cell} is negative. The reverse reaction is spontaneous.

$$E^\circ_{cell} = E^\circ_{cat} - E^\circ_{an} \text{ so } -0.44 \text{ V} = -0.44 \text{ V} - 0.00 \text{ V} = -0.44 \text{ V}$$

In some textbooks, the reversal of the reduction reaction is explained as an "oxidation potential," and the negative of the value of the standard reduction potential is used instead. In this case, the potentials are added together to find the cell potential. In this activity, we will use the difference in standard reduction potentials, as it is more widely used now than the oxidation potential. It is important to mention, however, because both conventions are still in use.

The more negative the standard reduction potential, the more strongly the product can transfer electrons to the H^+ ions in the SHE, and the more strongly it acts as a reducing agent.

The value of the standard cell potential is NOT affected by the number of electrons transferred in the overall redox reaction. However, take note of the following relationship: 1 *Joule* (J) of electrical energy is released by 1 *Coulomb* (C) of charge flowing through a potential difference of 1 *volt* (V). The sign and value of the standard reduction potential is related to the free energy change for the redox reaction taking place; a simple equation that must take into account the number of electrons transferred in the balanced redox reaction, *n*, and the charge of one mole of electrons, which is 96,485 C/mole. This value is known as *Faraday's constant*, symbolized as *F*. The complete equation is

$$\Delta G^\circ = -nFE^\circ_{cell}$$

Table 1   **Standard Reduction Potentials**

Cathode (Reduction) Half-Reaction	Standard Potential E° (volts)
$Li^+ (aq) + e^- \rightarrow Li (s)$	−3.04
$K^+ (aq) + e^- \rightarrow K (s)$	−2.92
$Ca^{2+} (aq) + 2e^- \rightarrow Ca (s)$	−2.76
$Na^+ (aq) + e^- \rightarrow Na (s)$	−2.71
$Mg^{2+} (aq) + 2e^- \rightarrow Mg (s)$	−2.38
$Al^{3+} (aq) + 3e^- \rightarrow Al (s)$	−1.66
$2H_2O (l) + 2e^- \rightarrow H_2 (g) + 2OH^- (aq)$	−0.83
$Zn^{2+} (aq) + 2e^- \rightarrow Zn (s)$	−0.76
$Cr^{3+} (aq) + 3e^- \rightarrow Cr (s)$	−0.74
$Fe^{2+} (aq) + 2e^- \rightarrow Fe (s)$	−0.41
$Cd^{2+} (aq) + 2e^- \rightarrow Cd (s)$	−0.40
$Ni^{2+} (aq) + 2e^- \rightarrow Ni (s)$	−0.23
$Sn^{2+} (aq) + 2e^- \rightarrow Sn (s)$	−0.14

Cathode (Reduction) Half-Reaction	Standard Potential E° (volts)
Pb^{2+} (aq) + 2e$^-$ $\longrightarrow$ Pb (s)	−0.13
Fe^{3+} (aq) + 3e$^-$ $\longrightarrow$ Fe (s)	−0.04
$2H^+$ (aq) + 2e$^-$ $\longrightarrow$ H_2 (g)	0.00
Sn^{4+} (aq) + 2e$^-$ $\longrightarrow$ Sn^{2+} (aq)	0.15
Cu^{2+} (aq) + e$^-$ $\longrightarrow$ Cu^+ (aq)	0.16
ClO_4^- (aq) + H_2O (l) + 2e$^-$ $\longrightarrow$ ClO_3^- (aq) + $2OH^-$ (aq)	0.17
AgCl (s) + e$^-$ $\longrightarrow$ Ag (s) + Cl$^-$ (aq)	0.22
Cu^{2+} (aq) + 2e$^-$ $\longrightarrow$ Cu (s)	0.34
ClO_3^- (aq) + H_2O (l) + 2e$^-$ $\longrightarrow$ ClO_2^- (aq) + $2OH^-$ (aq)	0.35
IO^- (aq) + H_2O (l) + 2e$^-$ $\longrightarrow$ I– (aq) + $2OH^-$ (aq)	0.49
Cu^+ (aq) + e$^-$ $\longrightarrow$ Cu (s)	0.52
I_2 (s) + 2e$^-$ $\longrightarrow$ 2I$^-$ (aq)	0.54
ClO_2^- (aq) + H_2O (l) + 2e$^-$ $\longrightarrow$ ClO^- (aq) + $2OH^-$ (aq)	0.59
Fe^{3+} (aq) + e$^-$ $\longrightarrow$ Fe^{2+} (aq)	0.77
Hg_2^{2+} (aq) + 2e$^-$ $\longrightarrow$ 2Hg (l)	0.80
Ag+ (aq) + e$^-$ $\longrightarrow$ Ag (s)	0.80
Hg^{2+} (aq) + 2e$^-$ $\longrightarrow$ Hg (l)	0.85
ClO^- (aq) + H_2O (l) + 2e$^-$ $\longrightarrow$ Cl$^-$ (aq) + $2OH^-$ (aq)	0.90
$2Hg^{2+}$ (aq) + 2e$^-$ $\longrightarrow$ Hg_2^{2+} (aq)	0.90
NO_3^- (aq) + $4H^+$ (aq) + 3e$^-$ $\longrightarrow$ NO (g) + $2H_2O$ (l)	0.96
Br_2 (l) + 2e$^-$ $\longrightarrow$ 2Br$^-$ (aq)	1.07
O_2 (g) + 4H+ (aq) + 4e$^-$ $\longrightarrow$ $2H_2O$ (l)	1.23
$Cr_2O_7^{2-}$ (aq) + $14H^+$ (aq) + 6e$^-$ $\longrightarrow$ $2Cr^{3+}$ (aq) + $7H_2O$ (l)	1.33
Cl_2 (g) + 2e$^-$ $\longrightarrow$ 2Cl$^-$ (aq)	1.36
Ce^{4+} (aq) + e$^-$ $\longrightarrow$ Ce^{3+} (aq)	1.44
MnO_4– (aq) + 8H+ (aq) + 5e$^-$ $\longrightarrow$ Mn^{2+} (aq) + $4H_2O$ (l)	1.49
H_2O_2 (aq) + 2H+ (aq) + 2e$^-$ $\longrightarrow$ $2H_2O$ (l)	1.78
Co^{3+} (aq) + e$^-$ $\longrightarrow$ Co^{2+} (aq)	1.82
$S_2O_8^{2-}$ (aq) + 2e$^-$ $\longrightarrow$ $2SO_4^{2-}$ (aq)	2.01
O_3 (g) + 2H+ (aq) + 2e$^-$ $\longrightarrow$ O_2 (g) + H_2O (l)	2.07
F_2 (g) + 2e$^-$ $\longrightarrow$ 2F$^-$ (aq)	2.87

| MODEL 1 | STANDARD REDUCTION POTENTIALS AND ASSOCIATED STANDARD FREE ENERGY CHANGES |

Positive Standard Reduction Potentials

$$Cu^{2+}(aq) + 2e^- \rightarrow Cu(s) \quad E^\circ_{red} = +0.34 \text{ V } (cathode\ reaction\ as\ written)$$

$$H_2(g) \rightarrow 2H^+(aq) + 2e^- \quad E^\circ_{red} = 0.00 \text{ V } (anode\ reaction,\ reversed)$$

$$Cu^{2+}(aq) + H_2(g) \rightarrow Cu(s) + 2H^+(aq) \quad (overall\ reaction)$$

$$E^\circ_{cell} = E^\circ_{cat} - E^\circ_{an} \quad = +0.34 \text{ V} - 0.00 \text{ V}$$
$$= +0.34 \text{ V } (spontaneous,\ positive\ value)$$

$$\Delta G^\circ = -nFE^\circ_{cell} = -(2\ mol\ e^-)\left(\frac{96,485\ C}{mol}\right)(+0.34\ V)\left(\frac{1\ J}{CV}\right)\left(\frac{1\ kJ}{1000\ J}\right)$$

$$\Delta G^\circ = -65.6 \text{ kJ } (spontaneous,\ negative\ value)$$

Negative Standard Reduction Potentials

$$Fe^{2+}(aq) + 2e^- \rightarrow Fe(s) \quad E^\circ_{red} = -0.41 \text{ V } (cathode\ reaction\ as\ written)$$

$$H_2(g) \rightarrow 2H^+(aq) + 2e^- \quad E^\circ_{red} = 0.00 \text{ V } (anode\ reaction,\ reversed)$$

$$Fe^{2+}(aq) + H_2(g) \rightarrow Fe(s) + 2H^+(aq) \quad (overall\ reaction)$$

$$E^\circ_{cell} = E^\circ_{cat} - E^\circ_{an} \quad = -0.41 \text{ V} - 0.00 \text{ V}$$
$$= -0.41 \text{ V}$$

The reaction as written above is non-spontaneous, as indicated by the negative cell potential, so the reverse reaction is spontaneous:

$$Fe(s) + 2H^+(aq) \rightarrow Fe^{2+}(aq) + H_2(g)$$

For the non-spontaneous reaction:

$$\Delta G^\circ = -nFE^\circ_{cell} = -(2\ mol\ e^-)\left(\frac{96,485\ C}{mol}\right)(-0.41\ V)\left(\frac{1\ J}{CV}\right)\left(\frac{1\ kJ}{1000\ J}\right)$$

$$\Delta G^\circ = +79.1 \text{ kJ } (non\text{-}spontaneous,\ positive\ value)$$

KEY QUESTIONS

1. In **Model 1**, what reaction is defined as the anode reaction in each case?

2. In **Model 1**, what is the defined potential of the anode reaction in each case?

3. Which overall redox reaction is non-spontaneous as written?

4. Write the spontaneous version of the reaction given in Key Question 3.

| MODEL 2 | CELL POTENTIALS AND STANDARD FREE ENERGY CHANGES |

Let's examine some of the Implications of Standard Reduction Potentials (Table 1)

Negative values of $E°_{red}$ are listed at the *top* of the table, where the reduced products are strong reducing agents, and their corresponding reactants are inert.

Positive values of $E°_{red}$ are listed at the *bottom* of the table, where the reactants are strong oxidizing agents, and the corresponding products are inert.

Spontaneous electron transfer takes place between reduced products, with more negative values of $E°_{red}$, and oxidized reactants, with more positive values of $E°_{red}$.

The arrow shows the spontaneous flow of electrons from the higher potential energy reduced reactant to lower potential energy oxidized reactant.

Cathode (Reduction Reaction)	Half-Reaction Standard Reduction Potential $E°$ (volts), vs. SHE
(Anode reaction, must be reversed)	
$Li^+(aq) + e^- \rightarrow Li(s)$	−3.04
$Fe^{2+}(aq) + 2e^- \rightarrow Fe(s)$	−0.41
$2H^+(aq) + 2e^- \rightarrow H_2(g)$	0.00
$Cu^{2+}(aq) + 2e^- \rightarrow Cu(s)$	+0.34
$F_2(g) + 2e^- \rightarrow 2F^-(aq)$	+2.87
(Cathode reaction, used as written)	

The balanced spontaneous overall reaction implied for this combination of reducing agent and oxidizing agent is:

$$Fe(s) + Cu^{2+}(aq) \rightarrow Fe^{2+}(aq) + Cu(s)$$

If the reaction is balanced differently, the number of electrons transferred, n, does not change the standard cell potential. However, it does change the standard free energy change, as calculated below.

$$E°_{cell} = E°_{cat} - E°_{an} = +0.34\ V - (-0.41\ V) = +0.75\ V \text{ (spontaneous, positive value)}$$

$$\Delta G° = -nFE°_{cell} = -(2\ mol\ e^-)\left(\frac{96,485\ C}{mol}\right)(+0.78\ V)\left(\frac{1\ J}{CV}\right)\left(\frac{1\ kJ}{1000\ J}\right)$$

$$\Delta G° = -145\ kJ \text{ (spontaneous, a negative value)}$$

INFORMATION

Any combination of half-reactions that involves a reduced species (product) having a more negative standard reduction potential and an oxidized species (reactant) having a more positive reduction potential, as in Table 1, will result in a spontaneous (positive) cell potential. When calculating the overall reaction, the reverse of the reaction with the more negative potential must be used, and the number of electrons on both sides must cancel.

Electrochemical cells that produce a electric current spontaneously with a positive standard cell potential are called *voltaic cells* or *galvanic cells*.

KEY QUESTIONS

5. In **Model 2**, which reaction takes place at the anode: oxidation or reduction? Identify the species that is oxidized or reduced.

6. Which reaction takes place at the cathode: oxidation or reduction? Identify the species that is oxidized or reduced.

7. Is the Fe metal an oxidizing agent or a reducing agent?

8. Which way do electrons flow through the wire: from the anode to the cathode, or from the cathode to the anode? Explain why this is the case.

9. Since electrons flow from the negative electrode to the positive electrode, which is positive: the Fe metal, or the Cu metal?

10. According to **Model 2**, how is the standard cell potential calculated from the standard half-cell potentials?

11. According to **Model 2**, how are the free energy change and the cell potential related?

12. Is it essential for the cell reaction in a voltaic cell to be spontaneous? Explain.

13. For the cell reaction to be spontaneous, what must be true of the free energy change and the cell potential?

EXERCISES

The following two half-reactions are used in a voltaic cell. A strip of cadmium metal is placed in a 1 M solution of cadmium sulfate, and a strip of copper is placed in a 1 M solution of copper sulfate. The metal strips are connected by a wire, and the solutions are connected by a salt bridge. Use this information to complete Exercises 1 through 7.

$$Cd^{2+} + 2e^- \rightarrow Cd \qquad E^\circ_{red} = -0.40 \text{ V}$$
$$Cu^{2+} + 2e^- \rightarrow Cu \qquad E^\circ_{red} = +0.34 \text{ V}$$

1. Identify the reducing agent and oxidizing agent from the species in these reactions.

2. Write and label the reactions occurring at the anode and at the cathode as "oxidation" or "reduction."

    anode:

    cathode:

3. Write the overall cell reaction.

4. Identify the species being oxidized and the species being reduced.

5. Identify the oxidizing agent and the reducing agent.

6. Calculate the standard cell potential $E°_{cell}$, and the standard free energy change $\Delta G°$, for the cell reaction.

7. One of the following does **not** take place in the Cd/Cu cell. Identify which.

   a. Electrons flow from the cadmium electrode to the copper electrode.

   b. The copper electrode increases in mass as the cell operates.

   c. Negative ions move through the salt bridge from the cadmium half-cell toward the copper half-cell.

   d. Cu^+ ions move through the salt bridge to the cadmium half-cell.

   e. Cu^{2+} ions move from the copper half-cell toward the cadmium half-cell.

8. Label the voltaic cell below. Include the anode, the cathode, the direction of the flow of electrons, and the direction of the flow of ions. This is based on the nickel and copper half reactions and is similar to the diagram in the first information section of this activity.

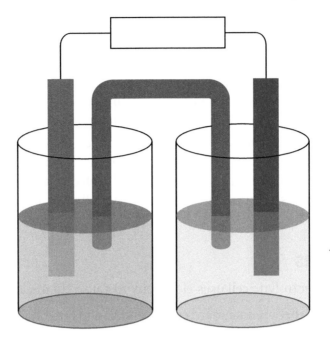

9. Using the table of standard cell potentials provided in this activity (Table 1), identify which of the following species is the strongest oxidizing agent and which is the weakest oxidizing agent.

$$Ni^{2+} \qquad Fe^{3+} \qquad I_2 \qquad O_2 \qquad Cr_2O_7^{2-} \qquad Cl_2$$

10. Using the table of standard cell potentials in Table 1, identify which of the following species is the strongest reducing agent and which is the weakest reducing agent.

$$Li \qquad Na \qquad Al \qquad Sn \qquad Cu \qquad I^- \qquad Br^-$$

Electrochemical Stoichiometry and the Nernst Equation

WHY?

Many redox reactions, and their corresponding electrochemical cells, are reversible (as you might predict from our daily use of rechargeable batteries). A cell that uses external electrical energy to produce a non-spontaneous chemical change is called an *electrolytic cell*. When working with electrolytic cells, you need to be able to identify the voltage and current necessary to operate a cell, as well as the chemicals needed for the anode and cathode compartments. Electrolysis is used to synthesize compounds, charge the battery in your car or cell phone, plate one metal with another (e.g., silver plated dinnerware), and produce aluminum metal, so understanding the relationship between the stoichiometry of the redox reaction and the electrical current used is also necessary. Stoichiometric factors are also required to determine the cell potential and equilibrium state under non-standard conditions.

LEARNING OBJECTIVES

- Understand how an electrolytic cell uses electricity to cause chemical reactions
- Recognize the difference between standard cell potentials and potentials at non-standard conditions.

SUCCESS CRITERIA

- Identify the components of an electrolytic cell
- Identify the chemical reactions that occur at each electrode
- Determine the amount of material produced or consumed by a cell and the electrical current used
- Calculate cell potentials under non-standard conditions

PREREQUISITES

15-3 *Free Energy and Chemical Equilibrium*

18-1 *Standard Reduction Potentials and Electrochemical Cells*

INFORMATION

Electrical current is measured in amperes (A), where 1 A = 1 C/s. 1 C (coulomb) is the charge on 6.2415×10^{18} electrons.

1 Faraday is the charge on 1 mole of electrons. 1 F = 96,485 C/mol e^-

Since an electrolytic cell is not spontaneous, the standard cell potential for an electrolytic cell, calculated from standard half-cell potentials, is negative. The opposite of this cell potential is the minimum voltage that must be applied to drive the reactions in the direction that is not spontaneous at standard conditions.

MODEL 1 · AN ELECTROLYTIC CELL FORCES A NON-SPONTANEOUS REDOX REACTION

Reconsider the voltaic cell in Activity 18-1:

$$Fe(s) + Cu^{2+}(aq) \rightarrow Fe^{2+}(aq) + Cu(s)$$

$$E°_{cell} = E°_{cat} - E°_{an} = +0.34\ V - (-0.41\ V) = +0.75\ V \text{ (spontaneous, positive value)}$$

$$\Delta G° = -nFE°_{cell} = -(2\ mol\ e^-)\left(\frac{96,485\ C}{mol}\right)(+0.75\ V)\left(\frac{1\ J}{CV}\right)\left(\frac{1\ kJ}{1000\ J}\right)$$

$$\Delta G° = -144\ kJ \text{ (spontaneous, a negative value)}$$

The two simple half-reactions involved are reversible, so the reverse reaction, which is non-spontaneous, can be driven by the application of an external voltage that is greater than −0.75 V to produce solid iron and copper(II) ions.

$$Fe^{2+}(aq) + Cu(s) \rightarrow Fe(s) + Cu^{2+}(aq)$$

Calculation showing that a 7.50 A current deposits 4.56 g of Fe in 35.0 min:

Figure 1

$$7.50\ A = 7.50\ C/s$$

$$7.50\ C/s \times 35.0\ min \times 60\ s/min = 15,750\ C$$

$$15,750\ C \times \left(\frac{1\ mole\ e^-}{96,485\ C}\right) = 0.1632\ mole\ e^-$$

$$0.1632\ mole\ e^- \times \left(\frac{1\ mole\ Fe}{2\ moles\ e^-}\right) = 0.08162\ mole\ Fe$$

$$0.08162\ mole\ Fe \times \left(\frac{55.845\ g}{mole}\right) = 4.56\ g\ Fe$$

KEY QUESTIONS

1. What are five or more important differences between the Fe/Cu voltaic cell and the Fe/Cu electrolytic cell illustrated in the models in this activity and the previous one?

2. According to **Model 1**, what is the minimum voltage that is needed to drive the reactions and run the cell under standard conditions? How is this value determined from standard reduction potentials for the half-reactions?

3. Why do the Cu/Fe voltaic and electrolytic cells have cell potentials with the same magnitude (0.75 V) but different signs (one is positive and the other negative)?

4. Is the free energy change for the electrolytic cell reaction positive or negative? Explain why, in terms of the cell potential and whether or not the reaction is spontaneous.

5. A calculation in **Model 1** shows that an electrical current of 7.50 A deposits 4.56 g of Fe in 35.0 min. Using this calculation, answer:

   a. How many moles of electrons flowed through the cell in 35 minutes?

   b. How many moles of zinc were plated onto the cathode in 35 minutes?

   c. Why is the ratio (moles of electrons) / (moles of Fe) = 2?

6. Name an item you use that might have been produced through electrolysis.

EXERCISES

1. The electrolysis of water involves the following two half-reactions.

$$2\,H_2O + 2e^- \rightarrow H_2 + 2\,OH^- \qquad\qquad 2\,H_2O \rightarrow O_2 + 4\,H^+ + 4e^-$$

a. Write the overall balanced equation for the cell reaction. *Note:* electrons cannot appear in the overall reaction equation.

b. Identify the species being oxidized and the species being reduced.

c. Calculate the standard cell potential using a table of standard reduction potentials.

d. Calculate the change in free energy under standard conditions.

2. Determine how long it will take to plate out 1.50 kg of aluminum in an electrolytic cell with a current of 240 A.

INFORMATION

In the electrochemical cells examined so far, standard conditions of reactant and product cation concentrations of 1.0 M and 25 °C are required. If gases are involved, as in the standard hydrogen electrode, they have to be at standard conditions (1.0 atm). From your previous study of thermodynamics and equilibrium, however, you know that the free energy of the system at non-standard conditions is given by the equation:

$$\Delta G_T = \Delta G_T^\circ + RT \ln(Q)$$

Since the relationship between cell potential and free energy change has been established as:

$$\Delta G^\circ = -nFE_{cell}^\circ$$

it is possible to rewrite the equation for free energy change at non-standard conditions in terms of cell potentials:

$$E_{cell} = E_{cell}^\circ - (RT/nF) \ln(Q)$$

This is known as the *Nernst equation*. If the concentrations of the reactant and product ions are 1 M, the ln(Q) term is equal to zero, and the observed value of E_{cell} is equal to E_{cell}°.

| MODEL 2 | THE POTENTIAL OF AN ELECTROCHEMICAL CELL UNDER NON-STANDARD CONDITIONS |

Reconsider the voltaic cell from Activity 18-1:

$$Fe(s) + Cu^{2+}(aq) \longrightarrow Fe^{2+}(aq) + Cu(s)$$

$$E_{cell}^\circ = E_{cat}^\circ - E_{an}^\circ = +0.337 \text{ V} - (-0.44 \text{ V}) = +0.78 \text{ V}$$

As the concentrations of reactants and products change over the course of the reaction, the observed value of E_{cell} changes accordingly. For the Fe/Cu cell, the concentration of Fe^{2+} ion increases as the concentration of Cu^{2+} ion decreases. If the concentration of Fe^{2+} ion increased to 1.99 M as the concentration of Cu^{2+} ion decreased to 0.01 M, the resulting potential could be calculated using these non-standard concentrations:

$$E_{cell} = E_{cell}^\circ - (RT/nF) \ln(Q)$$

$$E_{cell} = +0.78 \text{ V} - \left[\frac{\left(\dfrac{8.314 \text{ J}}{\text{K} \bullet mol}\right)(298 \text{ K})}{(2 \; mol \; e^-)\left(\dfrac{96,485 \text{ C}}{mol \; e^-}\right)\left(\dfrac{1 \text{ J}}{\text{CV}}\right)} \right] \ln\left(\frac{1.99}{0.01}\right)$$

$$E_{cell} = +0.78 \text{ V} - 0.07 \text{ V} = 0.71 \text{ V}$$

When the cell reaches equilibrium, $Q = K$, and $E_{cell} = 0$, so the following equation is also derived.

$$E_{cell} = E°_{cell} - (RT/nF) \ln(Q)$$

$$0 = E°_{cell} - (RT/nF) \ln(K)$$

$$E°_{cell} = (RT/nF) \ln(K)$$

KEY QUESTIONS

7. Why is the value of E_{cell} equal to 0 V at equilibrium?

8. Why is it necessary to use the conversion factor (1 J/CV) in the denominator of the RT/nF term of the Nernst equation?

EXERCISES

3. Derive an equation to calculate the value of the equilibrium constant, K, for a voltaic electrochemical cell.

4. If a two-electron biochemical reaction in the electron transport chain has a cell potential of 0.36 V, what is the value of the of the equilibrium constant for the reaction at 37 °C?

19-1 Coordination Compounds: An Introduction

Why?

Transition metals form unique compounds called *coordination compounds*. You need to know the characteristics of these compounds because they are vital to biological process and are widely used in industry, technology, and medicine. Some example applications of transition metals include jewelry, steel, paints, photographic films, magnetic tapes, anticancer drugs, and contrast agents in magnetic resonance imaging (MRI).

Success Criteria

- Identify the components and structures of coordination compounds
- Identify the geometry of coordination compounds
- Explain the bonding that takes place in coordination compounds

Prerequisites

05-3 *Electron Transfer Reactions*

07-4 *Multi-electron Atoms, the Aufbau Principle, and the Periodic Table*

08-2 *Lewis Model of Electronic Structure*

08-3 *Valence Shell Electron Pair Repulsion Model*

09-1 *Hybridization of Atomic Orbitals*

15-2 *The Reaction Quotient & Equilibrium Constant*

Information

Acid-base reactions have several definitions, each of which is used to emphasize different features of chemical reactivity. We have previously focused on the Brønsted-Lowry definition that deals with proton donors and acceptors. A *Lewis base* is defined as an electron pair donor. A *Lewis acid* is an electron pair acceptor. In many cases, the reaction between a Lewis acid and a Lewis base results in a relatively weak *coordinate covalent* bond. Consider the bond between boron trifluoride, BF_3, and ammonia, NH_3. BF_3 is electron deficient, because the octet rule is not satisfied for the central boron atom, and NH_3 has a nonbonding pair of electrons; this nonbonding pair can be used to form a bond with BF_3, yileding $F_3B - NH_3$. In this example, ammonia is the Lewis base, and boron trifluoride is the Lewis acid.

Metal ions often function as Lewis acids, accepting pairs of electrons from molecules or ions with lone pairs that act as Lewis bases. The resulting molecules or ions are called *coordination compounds*, and the molecules or ions surrounding the metal ion are called *ligands*. The number of coordinate covalent bonds formed by a compound is referred to as its *coordination number*. Ligands that donate only one pair of electrons, like water, ammonia, or the chloride ion, are called *mono-dentate* ligands. Many ligands can form more than one coordinate covalent bond; these ligands are called *poly-dentate* (or, more specifically, bi-, tri-, tetra-, etc.).

To summarize:

- A *ligand* is a molecular or anionic species that has atoms with nonbonding electrons available for donation to a metal

- A *polydentate ligand* contains more than one atom with nonbonding electrons available for donation. Similarly, a ligand with only one atom that can donate electrons is referred to as *monodentate*; one with two such atoms is *bidentate*, etc. Examples are shown in Figure 1.

- A *chelating agent* is a polydentate ligand which, when bonded to a metal, results in a complex ion with one or more rings, as shown in Figure 2

- A *coordination compound* or *complex* is a Lewis acid-base complex that contains a transition metal ion and several ligands covalently bonded to it. Such a complex can be cationic, anionic, or neutral.

- A *coordinate covalent bond* is a bond in which the shared electrons are contributed by one of the two bonded atoms. Once formed, the resulting bond is no different from a covalent bond in which both atoms contribute electrons. Arrows in Figure 3 (on the following page) indicate that the N atom in NH_3 uses a lone pair to bond to the metal.

- A *coordination number* is the number of atoms bonded to the transition metal in a coordination complex.

- *Counter ions* in a coordination compound serve to neutralize the charge on a complex ion

Figure 1 Mono- and bidentate ligands

Figure 2 Polydentate ligand and its complex

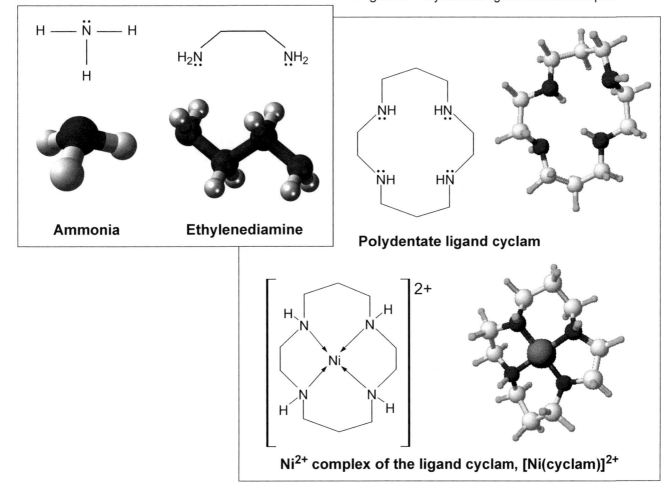

Ammonia   Ethylenediamine

Polydentate ligand cyclam

Ni^{2+} complex of the ligand cyclam, $[Ni(cyclam)]^{2+}$

MODEL   STRUCTURES OF COORDINATION COMPOUNDS

Abbreviations:

CN = coordination number

Ox = oxidation state of central metal

Charge = charge on the complex ion

Coordination compound	Complex ion	Geometry	Ligand(s)	Counter ion	CN	Ox	Charge
$[Co(NH_3)_6]Cl_3$	$[Co(NH_3)_6]^{3+}$	Octahedral	NH_3	Cl^-	6	+3	+3
$K_3[Fe(CN)_6]$	$[Fe(CN)_6]^{3-}$	Octahedral	CN^-	K^+	6	+3	−3
$Na_2[HgI_4]$	$[HgI_4]^{2-}$	Tetrahedral	I^-	Na^+	4	+2	−2
$[Pt(NH_3)_2Cl_2]$	None	Square planar	NH_3, Cl^-	None	4	+2	0
$K_2[PtCl_6]$	$[PtCl_6]^{2-}$	Octahedral	Cl^-	K^+	6	+4	−2

Figure 3 $[Co(NH_3)_6]^{3+}$ ion

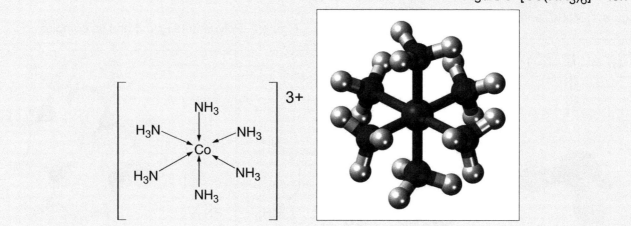

KEY QUESTIONS

1. What are the components that must be present in all coordination compounds?

2. What additional component may be present in some coordination compounds?

3. Why are the square brackets included in the chemical formula for a coordination compound?

4. What information is provided by a compound's coordination number?

5. When a coordination compound dissolves in water, it dissociates into a complex ion and its counter ions. What are the dissociation products of each coordination compound in the table?

6. How can the charge on a complex ion be calculated from other information in the **Model**?

7. How can the oxidation number of a transition metal be determined from other information in the **Model**?

8. What do the Lewis electron structures of the ligands in the table all have in common? How does this characteristic contribute to the formation of a coordination compound, in terms of the Lewis acid-base model?

EXERCISES

1. Sketch $K_3[Fe(CN)_6]$, showing its octahedral geometry and the location of the ligands and counter ions. Label the bonds appropriately as covalent, coordinate covalent, or ionic.

2. Sketch the complex ion $([Fe(CN)_6]^{3+})$, showing the **sp^3**-orbitals of the ligands overlapping with the **sp^{3}d^2** orbitals of the transition metal ion.

3. Sketch the trigonal planar and trigonal pyramidal geometries of complex ions with coordination number 3.

4. Illustrate the difference between the structures of complex ions containing Fe^{3+} bonded to the polydentate ligand, ethylenediamine (Figure 2), and to the monodentate ligand, NH_3. In both cases, the coordination number of Fe is 6. Remember to include the charge on the complex ion.

5. Using your experience with VSEPR theory, construct a table listing possible coordination numbers (2–6) and resulting geometries for complex ions.

19-2 Coordination Compounds: Magnetism and Color

WHY?

Coordination compounds and gemstones exhibit characteristic magnetic and color properties. These properties arise from the energy splitting of the transition metal's d-orbitals, caused by the presence of ligands in coordination compounds or anions in crystal lattices. It is important that you learn to recognize which d-orbitals are occupied by electrons, and how this occupation determines the magnetic properties and color of the compound. These properties give information about the bonding in coordination compounds and allow for the creation of compounds with useful magnetic and color characteristics.

LEARNING OBJECTIVES

• Understand the crystal field and ligand field model of d-orbital splitting

• Understand magnetism and color in coordination compounds

SUCCESS CRITERIA

• Draw d-orbital energy level diagrams for octahedral coordination complexes

• Identify whether transition metals are low spin or high spin

• Identify whether a coordination compound is diamagnetic or paramagnetic

• Relate the magnitude of the ligand field splitting to the color of the compound

PREREQUISITES

07-1 *Electromagnetic Radiation*

09-3 *Molecular Orbital Theory*

19-1 *Coordination Compounds: An Introduction*

INFORMATION

In colored gemstones like rubies, amethyst and topaz, trace amounts of transition metal ions are substituted into the crystal lattices of otherwise colorless minerals, in sites generally occupied by other cations. The colorless mineral corundum, Al_2O_3, is converted into ruby (red) when Cr^{3+} ions are present, substituting Mn^{3+} for the Al^{3+} cations yields amethyst (purple), and similarly substituting Fe^{3+} ions gives topaz (yellow). The early explanation for the colors of these gemstones is called *crystal field theory* (CFT). In that model, the effect of the surrounding oxide anions, which form "octahedral holes" occupied by the metal cations, is used to explain the colors. According to the crystal field model, d-orbitals in the transition metals are split from their usual degeneracy according to the differences in electrostatic repulsion between d-orbital electrons with the anions in the lattice. Those differences arise from the shapes of the d-orbitals. This simple model, which considers only electrostatic repulsion and the shapes of the d-orbitals, was later found to have significant similarities to the d-orbital splitting observed in coordinate compounds. In *ligand field theory* (LFT), bonding between the metal ions and ligands is fully considered. However, the d-orbital splitting that occurs in crystals also takes place in metal complexes. The LFT model describes the bonding, non-bonding, and anti-bonding molecular orbitals formed between the metal ion and the ligands. This activity will focus on the d-orbital splitting which is common to both theories, as it is this effect that explains the color and magnetism of coordination compounds.

According to the ligand field model, the splitting of the energies of the d-orbitals is a function of the relative positions of the ligands and the individual d-orbitals.

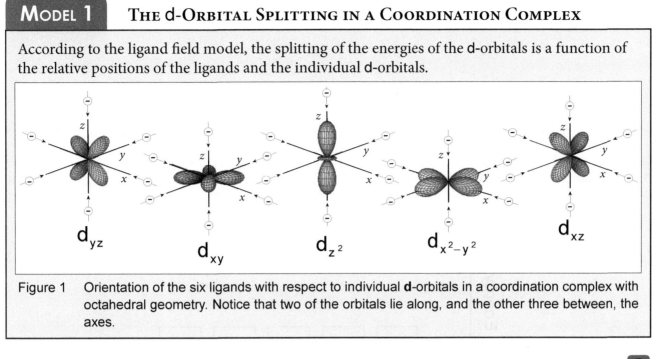

Figure 1   Orientation of the six ligands with respect to individual d-orbitals in a coordination complex with octahedral geometry. Notice that two of the orbitals lie along, and the other three between, the axes.

KEY QUESTIONS

1. Which d-orbitals lie along the x, y, or z-axes, and which lie between any two axes?

2. Which orbitals experience a direct interaction with the ligands that also lie along the axes?

3. Which orbitals do not experience a direct interaction with the ligands?

EXERCISES

1. Based on the nature of the interactions that you identified in Key Questions 2 and 3, divide the five d-orbitals into two sets. Explain the reasoning for your choices.

Set 1	Set 2

2. In which set of orbitals will the stronger interactions with the ligands be found: Set 1 or Set 2? Explain the reason for your choice (hint: consider the electrons).

3. The diagram below shows all five d-orbitals in a "free metal ion" (i.e., in the absence of ligands) at the same energy level:

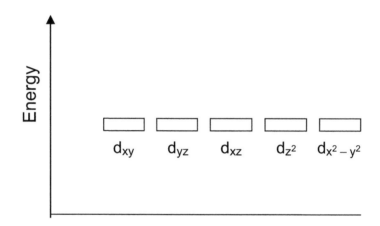

In crystal field theory, if a metal ion is in an octahedral hole in a crystal lattice with anions along the x, y, and z axes, the energies of two sets of d-orbitals depend on the extent of overlap of these orbitals with the ions. The set experiencing stronger interactions due to their orientations along the axes will be less stable and at a higher potential energy because the electron-electron interactions are repulsive. The set of d-orbitals with weaker interactions with the ions have electron density between the axes and are more stable as a result. Draw a diagram below, illustrating the relative energies of the two sets that you identified and characterized in Exercises 1 and 2.

The diagram that you have drawn is called an *octahedral-field-splitting energy-level diagram*. The energy difference between the two sets is referred to as Δ_o ("o" refers to octahedral geometry) and is called the *crystal-field splitting* (in the CFT model). In the LFT model, the explanation is different; it is based on molecular orbitals that arise from the overlap of the d-orbitals with the ligand-orbitals. The lower energy orbitals, which have less interaction with the ligands due to their shapes, are non-bonding molecular orbitals. The higher energy orbitals, which have strong overlap with the ligand-orbitals, are the sigma anti-bonding, σ^*, molecular orbitals. The sigma bonding orbitals created by the strong overlap along the x, y, and z-axes are much lower in energy and filled with electrons, and do not contribute to the observed *octahedral-ligand-field splitting* in the LFT model.

MODEL 2 — EFFECT OF DIFFERENT LIGANDS ON THE MAGNITUDE OF Δ_o

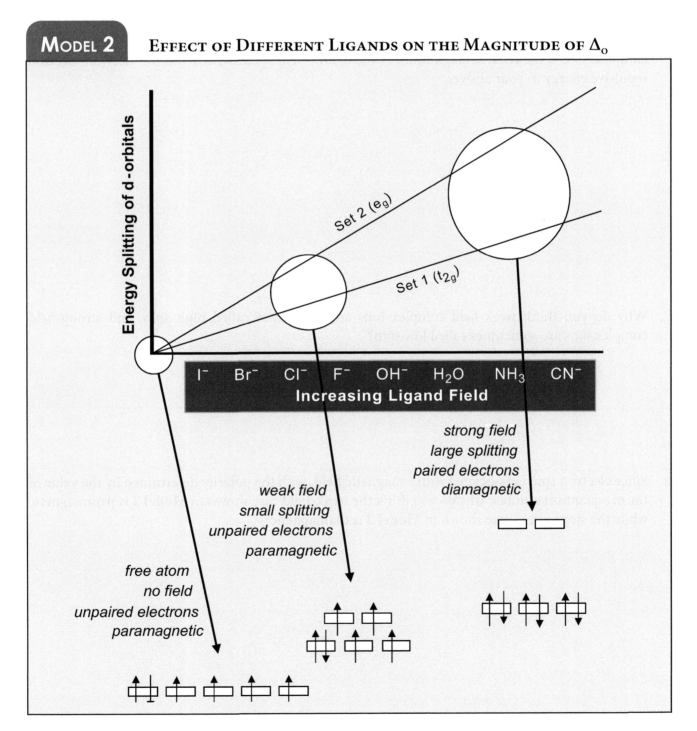

KEY QUESTIONS

4. Which ligand shown in **Model 2** has the largest crystal field effect (i.e., causes the greatest splitting of the d-orbital sets)?

5. What is the difference, as shown in **Model 2**, between the ways the electrons fill the d-orbitals in the weak-field complexes and in the strong-field complexes?

6. Since electrons fill orbitals to produce the lowest possible energy state, why do you think electrons are unpaired in the free atom and weak-field complexes, while they are paired-up in the strong-field complexes? Use the relative magnitudes of the ligand-field splitting, Δ_o, and the electron-electron repulsive energy in your answer.

7. Why do you think weak-field complex ions are sometimes called high spin, and strong-field complex ions are sometimes called low spin?

8. Since electron spin is associated with a magnetic field, with the polarity determined by the value of the m_s quantum number, why do you think the weak-field case shown in **Model 2** is paramagnetic while the strong field case shown in **Model 2** is diamagnetic?

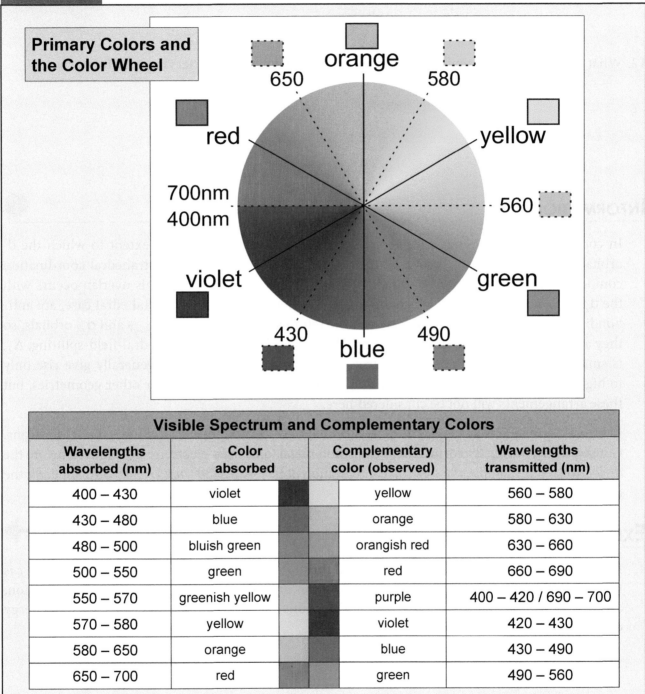

Primary Colors and the Color Wheel

Visible Spectrum and Complementary Colors			
Wavelengths absorbed (nm)	Color absorbed	Complementary color (observed)	Wavelengths transmitted (nm)
400 – 430	violet	yellow	560 – 580
430 – 480	blue	orange	580 – 630
480 – 500	bluish green	orangish red	630 – 660
500 – 550	green	red	660 – 690
550 – 570	greenish yellow	purple	400 – 420 / 690 – 700
570 – 580	yellow	violet	420 – 430
580 – 650	orange	blue	430 – 490
650 – 700	red	green	490 – 560

KEY QUESTIONS

9. A solution of $[CrF_6]^{3-}$ absorbs wavelengths in the range 610 – 690 nm. What is the color of the solution?

10. A solution of $[Cr(H_2O)_6]^{3+}$ is violet. What color and wavelengths are absorbed by this complex ion?

11. A solution of $[Cr(NH_3)_6]^{3+}$ absorbs wavelengths in the range of 435 to 480 nm. What color is this solution, and what color does it absorb?

12. What are examples that you have observed of magnetism and color serving a useful purpose?

INFORMATION

In coordination compounds with geometries other than octahedral, the extent to which the d-orbitals overlap with the geometry of the ligand field is different. For a tetrahedral coordination compound, the stronger overlap is between the ligands and d-orbitals. This overlap occurs with the d_{xy}, d_{xz}, and d_{yz} orbitals, which are higher in energy and, as in the octahedral case, are anti-bonding. The least overlap occurs between the tetrahedral field and the $d_{x^2-y^2}$ and d_{z^2} orbitals, so they are lower in energy and non-bonding. With fewer ligands, the tetrahedral-field-splitting, Δ_T, is smaller than Δ_o. Tetrahedral geometries in coordination compounds generally give rise only to high-spin cases. Other d-orbital splitting arrangements are observed for other geometries, but these arrangements will not be considered here.

In transition metals, electrons in the outer s-orbitals have a lower ionization energy than d-electrons. Consequently, during the formation of transition metal ions, the s-electrons are removed before the d-electrons. For example, the electron configuration of Fe^{2+} is $[Ar]3d^6$, *not* $[Ar]4s^23d^4$. Similarly, the electron configuration of Fe^{3+} is $[Ar]3d^5$.

EXERCISES

4. Recall that wavelength and energy are related through the equations $E = h\nu$ and $c = \nu\lambda$, where c is the speed of light, ν is the frequency of the radiation, and λ its wavelength. Using this information, arrange the chromium compounds in Key Questions 9, 10, and 11 in order of decreasing energy of the photons absorbed.

5. Examine the d-orbital energy level diagrams in **Model 2**, and draw similar diagrams for octahedral complex ions of each of the following:

Fe^{3+} (weak field) Fe^{3+} (strong field)

Ni^{2+} (weak field) Ni^{2+} (strong field)

6. Identify the number of unpaired electrons in each of the following complex ions, and determine whether the ion is diamagnetic or paramagnetic:

$$[Zn(NH_3)_4]^{2+} \quad [Fe(H_2O)_6]^{2+}$$

PROBLEMS

1. The octahedral complex of SCN^- with Fe^{3+} is paramagnetic with five unpaired electrons. Does SCN^- produce a stronger or weaker ligand field than CN^-? Explain.

2. The coordination compound $[Co(H_2O)_6]^{3+}$ is violet (absorbs yellow light). The coordination compound $[Co(CN)_6]^{3+}$ is yellow (absorbs violet light).

	$[Co(H_2O)_6]^{3+}$	$[Co(CN)_6]^{3+}$
a. Which compound absorbs light with the shorter wavelength?	☐	☐
b. Which compound has the smaller ligand field splitting of the d-orbitals?	☐	☐

    c. Which ligand produces the stronger field?

3. If you want to determine whether a given ligand is a strong-field or a weak-field ligand by calculating the number of unpaired electrons from the (experimentally determined) paramagnetism, would it be better to use octahedral complexes of Cr^{2+} or Ni^{2+}? Explain.

4. The electron in the d-orbital of $Ti(H_2O)_6^{3+}$ absorbs light at 570 nm and moves from a lower energy d-orbital, called t_2g, to a higher energy d-orbital, called e_g.

    a. How large is the ligand field splitting in this octahedral complex ion, in units of kJ/mol?

    b. How large is the ligand field splitting compared to the first ionization energy of titanium (659 kJ/mol)?

WHY?

A *functional group* is a specific arrangement of atoms in an organic molecule. Functional groups determine a compound's chemical reactivity and physical properties. Functional groups are important because they make it possible to arrange a very large number of organic molecules into a much smaller number of classes. Molecules in the same class (with the same functional groups) undergo the same types of chemical reactions and have similar physical properties. If you are able to recognize a molecule's functional groups, then you will be able to identify and understand its chemical reactivity and physical properties. The names of organic compounds are also directly connected to the functional groups they contain.

LEARNING OBJECTIVE

- Associate the arrangement of atoms in a molecule with specific functional groups

SUCCESS CRITERIA

- Identify a molecule's functional groups from its structure
- Draw a structure for a given functional group

PREREQUISITES

08-2   *Lewis Model of Electronic Structure*

03-2   *Representing Molecular Structures*

INFORMATION & REVIEW

In previous activities, distinctions were made between *molecular formulas, structural formulas, Lewis structures,* and *line-drawings.* A molecular formula simply shows the number of atoms of each element present in a molecule, without any information about its structure. For example, acetic acid has the molecular formula of $C_2H_4O_2$. Structural formulas are often used for organic compounds, since carbon atoms tend to bond with each other, and to hydrogen atoms, in order to complete their octets. The structural formula for acetic acid is CH_3COOH. The bonding in the –COOH *functional group* is not quite clear from the structural formula, but it can be determined and shown more clearly using a Lewis structure, shown here in Figure 1. Since functional groups appear as recurring bonding arrangements in different organic molecules, the structure implied by RCOOH is understood to represent the bonding in the Lewis structure, and R is used to represent the rest of the organic molecule. As you have also seen, line drawings can also be used to describe the bonding in an organic molecule even more simply, using lines to represent bonded carbon atoms. Carbon atoms are implied at each end of a line, and the appropriate number of hydrogen atoms to complete each carbon's octet are assumed to be present. Multiple bonds are shown with multiple lines, just as in a Lewis structure. Non-bonding electrons are not shown in line drawings. The line drawing for acetic acid is shown here in Figure 2. Note that hydrogen atoms in the functional groups are shown.

Figure 1   Figure 2

Contributed by Joseph Lauher, Stony Brook University

	MODEL	**SOME COMMON FUNCTIONAL GROUPS**

Name	Lewis Structure	General Formula	Example
alkane		RH	CH_4, methane CH_3CH_3, ethane
alkene		$R_2C=CR_2$	CH_2CH_2, ethene
alkyne	H—C≡C—H	RC≡CR	CHCH, ethyne
aromatic ring		ArH	C_6H_6, benzene
haloalkane		RX	CH_3Cl, chloromethane
alcohol		ROH	CH_3CH_2OH, ethanol
ether		ROR	CH_3OCH_3, dimethylether
amine		R_3N	CH_3NH_2, methylamine

Name	Structure	General Formula	Example
aldehyde	$H-\overset{\overset{H}{|}}{\underset{\underset{H}{|}}{C}}-\overset{\overset{H}{|}}{C}=\ddot{O} :$	R(CO)H	$CH_3(CO)H$, acetaldehyde
ketone	$H-\overset{\overset{H}{|}}{\underset{\underset{H}{|}}{C}}-\overset{}{\underset{\underset{:\ddot{O}:}{\parallel}}{C}}-\overset{\overset{H}{|}}{\underset{\underset{H}{|}}{C}}-H$	R_2CO	$(CH_3)_2CO$, acetone
carboxylic acid	$H-\overset{\overset{H}{|}}{\underset{\underset{H}{|}}{C}}-C\overset{\diagup\,\ddot{O}:}{\diagdown\,:\ddot{O}-H}$	RCOOH	CH_3COOH, acetic acid
ester	$H-\overset{\overset{H}{|}}{\underset{\underset{H}{|}}{C}}-\overset{}{\underset{\underset{:\ddot{O}:}{\parallel}}{C}}-\ddot{O}-\overset{\overset{H}{|}}{\underset{\underset{H}{|}}{C}}-H$	R(CO)OR	$CH_3(CO)OCH_3$, methyl acetate
amide	$H-\overset{\overset{H}{|}}{\underset{\underset{H}{|}}{C}}-\overset{}{\underset{\underset{:\ddot{O}:}{\parallel}}{C}}-\overset{\ddot{}}{\underset{\underset{H}{|}}{N}}-H$	$R(CO)NR_2$	$CH_3(CO)NH_2$, acetamide

TASK

In the table that follows, identify the distinguishing characteristics of each functional group listed in the model on the previous page(s).

alkane	
alkene	
alkyne	
aromatic ring	
haloalkane	

alcohol	
ether	
amine	
aldehyde	
ketone	
carboxylic acid	
ester	
amide	

EXERCISES

1. Identify the functional groups contained in the following compounds.

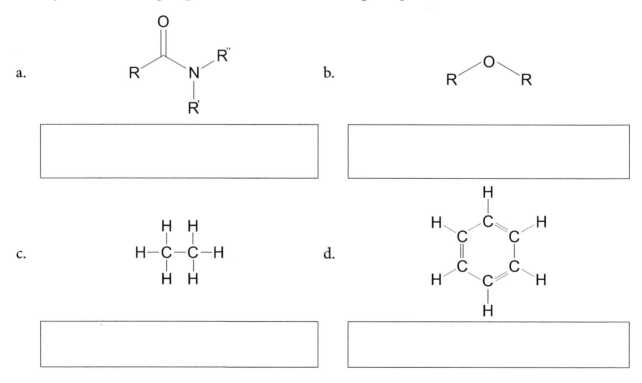

2. Draw the line structure of molecules containing the following functional groups.

a. alkyne

b. ether

c. aldehyde

d. ester

e. amide

3. Write the molecular formula for each of the following compounds, and identify all the structural groups in each.

a. aspartic acid, an amino acid found in asparagus (among other places)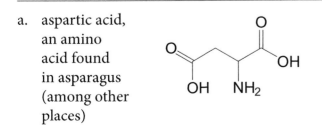

b. salicylic acid, found in acne medicine

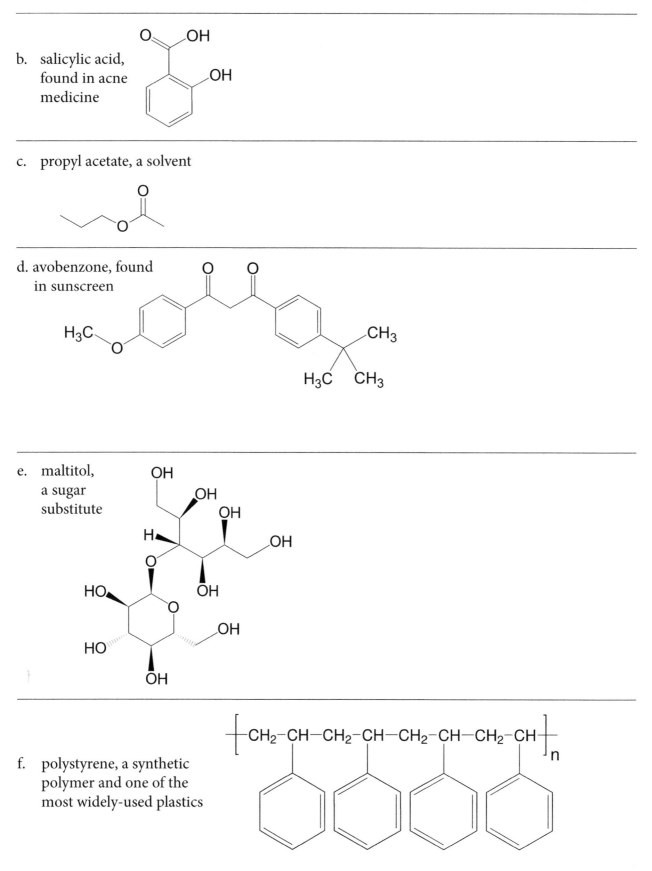

c. propyl acetate, a solvent

d. avobenzone, found in sunscreen

e. maltitol, a sugar substitute

f. polystyrene, a synthetic polymer and one of the most widely-used plastics

WHY?

Polymers are large molecules that are composed of repeating units called *monomers.* Because the properties of polymers are extremely diverse, they have widespread uses, more of which are discovered every day. They are key components in living organisms, and have many applications in daily life, medicine, technology, and industrial processes. You should know something about the properties of polymers and how polymers are made, since you encounter them daily, and the jobs of many chemists involve working with polymers.

LEARNING OBJECTIVES

- Identify reactions that are used to synthesize polymers
- Summarize information on polymer structure, properties, and uses

SUCCESS CRITERIA

- Correctly identify polymerization reactions and properties of polymers
- Determine both the reactants and products of polymerization reactions

PREREQUISITES

03-2   *Representing Molecular Structures*

05-2   *Introduction to Acid – Base Reactions*

INFORMATION

Addition Polymers

Small molecules with multiple bonds can come together as *monomers* to form larger molecules called *polymers.* During the polymerization process, the multiple bonds are broken and new, single bonds between the monomers are formed. Each monomer is added to the end of a growing chain, so these molecules are called *addition polymers.* The formation of polyethylene from ethylene is an example of addition polymerization.

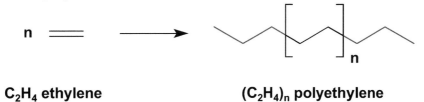

C_2H_4 ethylene $(C_2H_4)_n$ polyethylene

KEY QUESTIONS

1. What is the relationship between a monomer and a polymer?

2. What are the characteristics of addition polymerization?

EXERCISES

1. The following molecules also undergo addition polymerization via the ethylene double bond. Draw the respective polymers.

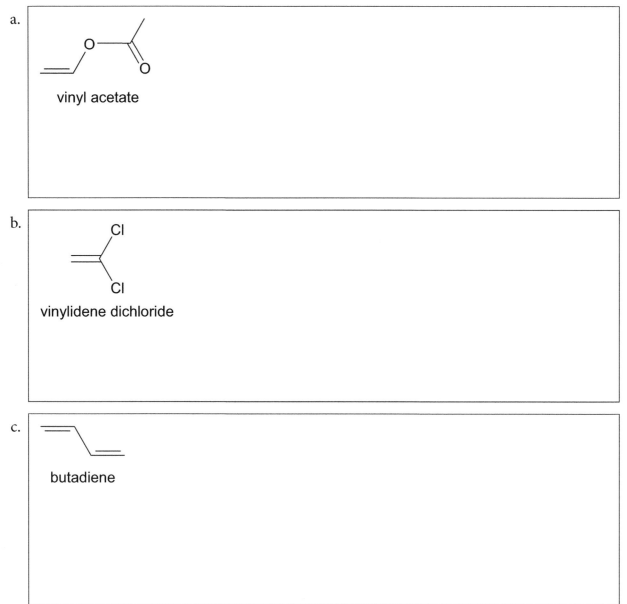

a.

vinyl acetate

b.

vinylidene dichloride

c.

butadiene

2. What monomers would you use to form the following addition polymers?

a.

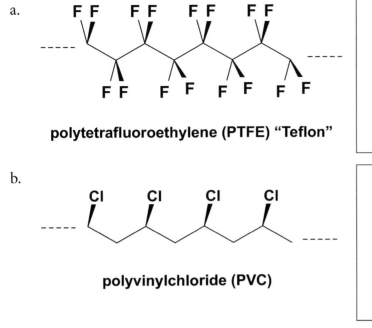

polytetrafluoroethylene (PTFE) "Teflon"

b.

polyvinylchloride (PVC)

PROBLEMS

1. When the molecule styrene is polymerized, a small amount of a compound called *divinylbenzene* is sometimes added. This addition makes the resulting polymer harder and stronger, via a process called *crosslinking*. Explain crosslinking in terms of addition reactions involving these two reactants. Illustrate your explanation with a drawing of the cross-linked polymer.

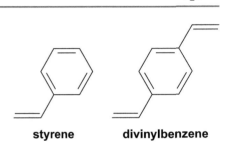

styrene   **divinylbenzene**

2. *Natural rubber* is a polymer of the isoprene molecule. When polymerized, it produces ***polyisoprene,*** a linear chain with a methyl side group next to a double bond in the chain.

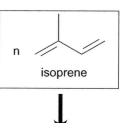

isoprene

a. Draw the polymer below. Label it, "natural rubber".

b. Charles Goodyear discovered a process called *vulcanization* that led to an increase in the strength and durability of rubber by crosslinking the polymer with the element sulfur. The double bonds in the polyisoprene chain are **not** affected by the cross linking. Draw the crosslinked polymer below, labeling it, "vulcanized natural rubber". (If you can't figure this out, search for pictures in your text or online.)

S_8 + heat

INFORMATION

Condensation Polymers

In a condensation polymer, monomers form bonds between their functional groups by eliminating a small molecule such as water or HCl. Two common examples are esters and amides. An ester is shown at right and an amide is found on the following page.

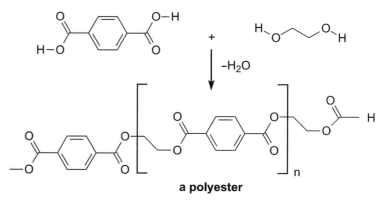

a polyester

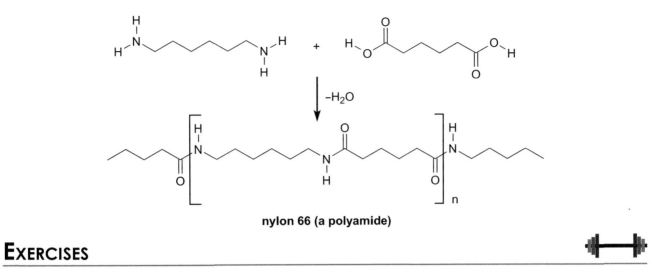

nylon 66 (a polyamide)

EXERCISES

3. Draw the polymers that would result from the following condensation reactions, and identify the small molecule that is eliminated in the reaction.

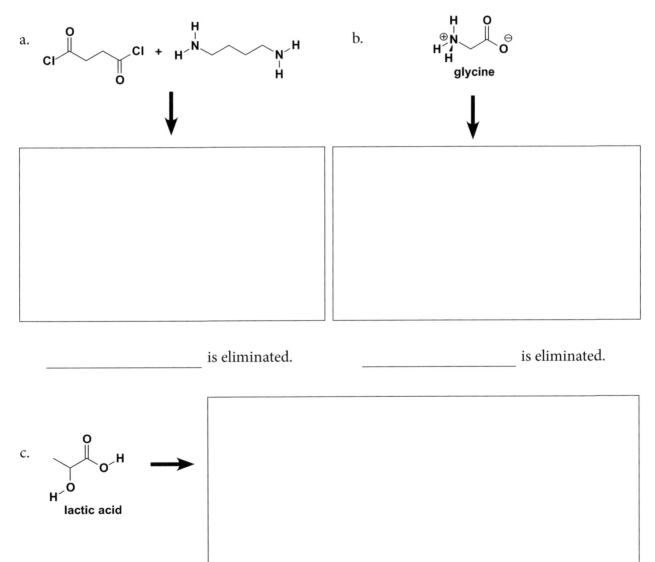

a.

b. glycine

_____ is eliminated.   _____ is eliminated.

c. lactic acid

_____ is eliminated.

4. What monomers would result in the following condensation polymers?

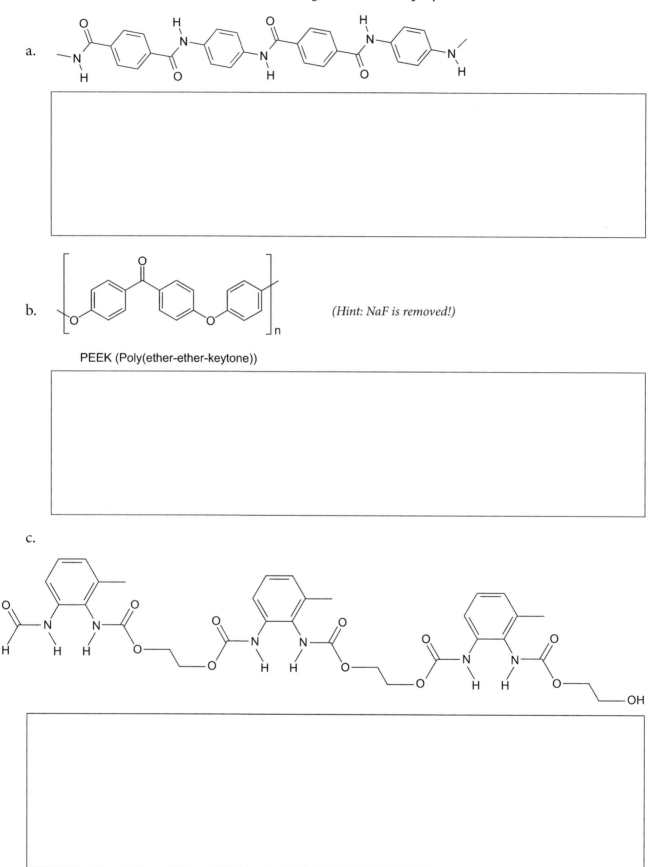

a.

b.

(Hint: NaF is removed!)

PEEK (Poly(ether-ether-keytone))

c.

3. What structural features must a molecule have in order to form a polymer in a condensation reaction?

4. What structural features must a molecule have in order to form a polymer in an addition reaction?

5. What features in the structure of a polymer would make the material very strong and rigid?

6. Why is it that, when heated, polymers do not melt and become liquid, in the same way that ice melts to become water?

ACTIVITY
21-1 Amino Acids and Proteins

WHY?

Proteins constitute about 50% of the dry weight of most cells, and each type of protein has its own unique structure and function. Knowledge of the structures of these molecules is essential for continuing studies in chemistry and biology.

LEARNING OBJECTIVES

- Understand how amino acids combine to form proteins
- Understand the different properties of amino acids

SUCCESS CRITERIA

- Identify the two major types of protein secondary structures: α-helices and β-sheets
- Correctly illustrate the structures of amino acids and peptides

PREREQUISITES

03-2   *Representing Molecular Structures*

20-2   *Polymer Chemistry*

INFORMATION

Polymers are any kind of large molecule or *macromolecule* made up of repeating identical or similar subunits called *monomers*. Amino acids are the monomers that are used to build proteins. To form proteins, the amino acids are linked by a condensation reaction that produces an amide linkage (-CO-NH-) and eliminates water. The amide linkage in proteins is also called a *peptide linkage*, and the chain of amino acids is also known as a *polypeptide*, or simply a *peptide*. Generally, peptides consist of no more than 50 to 100 amino acids, while most proteins typically consist of 100 to 1000 amino acids. Some proteins contain only one polypeptide chain, while others, such as hemoglobin, contain several polypeptide chains all twisted together. The sequence of amino acids in each polypeptide or protein is unique and consequently each protein has its own unique 3-D shape or *native conformation*.

If even one amino acid in the sequence is changed, the protein's ability to function can be impared. For example, sickle cell anemia is caused by a change in only one nucleotide in the DNA sequence, which causes just one amino acid in one of the hemoglobin polypeptide molecules to be different. Because of this, the whole red blood cell ends up deformed and is unable to carry oxygen properly.

The amino acids are known as *α-amino acids* because (with the exception of proline) they all have the same structure: a primary amino group and a carboxylic acid group attached to the same α-carbon that carries a side chain.

Contributed by Frank Fowler and Andisheh Abedini, Stony Brook University

All amino acids have the same general structural formula:

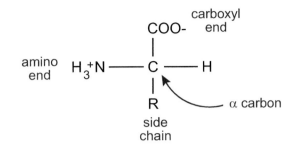

A protein's *primary structure* (1°) is the amino acid sequence of its polypeptide chain(s).

A protein's *secondary structure* (2°) is the local spatial arrangement of a polypeptide's backbone atoms, without regard to the *conformations* (possible spatial arrangements) of its side chains.

A protein's *tertiary structure* (3°) refers to the three-dimensional structure of an entire polypeptide. The tertiary structure is stabilized by the interactions between the individual side groups (R) on the amino acids comprising the protein. Many proteins are composed of two or more polypeptide chains, referred to as *subunits*, which associate through noncovalent interactions and, in some cases, disulfide bonds. A protein's *quaternary structure* (4°) refers to the spatial arrangement of these subunits.

KEY QUESTIONS

1. Which monomers are the building blocks of proteins?

2. Which type of polymerization reaction serves to attach monomers to each other, producing peptides and proteins?

3. What is the name of the linkage or bond formed between the monomers in a protein?

4. What is it about their structure that gives different α-amino acids different properties?

5. What is it about their structure that gives different proteins different properties?

6. If there are 20 different amino acids, how many different peptides can be produced with a chain length of 50 amino acids?

7. Why is it important to recognize that proteins have a primary, secondary, tertiary, and quaternary structure?

EXERCISES

1. The various amino acids are usually classified according to the polarities of their side chains, **R**, which are attached to the α-carbon. Use your text or another resource as a reference to complete the following exercises.

    a. Write the names and three-letter abbreviations for the amino acids with nonpolar side chains.

    b. Write the names and three-letter abbreviations for the amino acids with uncharged polar side chains.

    c. Write the names and three-letter abbreviations for the amino acids with charged polar side chains.

2. In the physiological pH range (around 7), both the carboxylic acid and the amino groups of the amino acids are completely ionized, meaning that they carry charges. These ions have both a positive and a negative charge and are called *zwitterions*, which is from the German term for hybrid ions. Draw the structure of an amino acid in the zwitterion form. Include the carboxylic acid group and the amino group, designate the side chain on the α-carbon as R, and label the location of the + and – charges.

3. Write the reaction equation, using structural formulas, that shows how alanine and glycine react to form the dipeptide alanylglycine.

4. Draw the structural formula of three peptide-bonded amino acids, using R1, R2, and R3 to designate the side chains.

5. Draw the following peptides in their predominant ionic forms at pH 7:

   a. Phe-Met-Arg

b.   Gln-Ile-His-Thr-Arg

c.   Gly-Pro-Tyr-Cys-Lys

1. One major type of protein secondary structure is the α-helix. α-helices are held together by hydrogen bonds and have a right-handed turn every 3.6 residues. The hydrogen bonds of α-helices are arranged such that the peptide C=O bond of the nth residue points along the helix, towards the peptide N-H group of the $(n + 4)^{th}$ residue. The R groups of the α-helices all project outward and downward from the helix to avoid steric interference with the polypeptide backbone and with each other.

   Draw an example of an α-helix. On the helix, show the number of amino acids per turn as squares; designate where hydrogen bonds would form between C=O and –NH. Also show the locations of the side chains. (***Hint:*** Find a figure in your text or other resource.)

2. β-sheets are the second major type of secondary structure. In a β-sheet, hydrogen bonding occurs between neighboring polypeptide chains, rather than within, as it does in α-helices.

   a. Use your text or another resource to help you identify the two major types of β-sheets, and explain the difference between the two.

b. Draw antiparallel and parallel β-sheets, using arrows to illustrate the direction of chains.

3. In your own words, define what protein tertiary and quaternary structure entails.

ACTIVITY 22-1 Nuclear Chemistry: Binding Energy

WHY?

Nuclear reactions involve changes in the structures of the nuclei of atoms. While nuclear reactions are, in some ways, similar to chemical reactions, there are important differences that must be taken into account. The very existence of elements other than hydrogen can only be explained in terms of nuclear reactions and the *binding energy* that allows protons and neutrons to combine to form other elements and isotopes. In this activity, some strongly held principles of Dalton's atomic theory, the Law of Conservation of Mass, and the Law of Conservation of Energy will have to be re-examined in light of nuclear reactions and binding energy.

WHAT DO YOU THINK?

Centuries ago, a major goal of alchemists was to turn lead into gold. Do you think this is possible? Explain why or why not.

LEARNING OBJECTIVE

- Understand the origin of the energy produced in nuclear reactions

SUCCESS CRITERIA

- Explain the origin of the energy released in a nuclear reaction
- Correctly calculate nuclear binding energies and the energy released in a nuclear reaction

PREREQUISITES

INFORMATION

Chemical Structure and Reactions

Chemical structure, intermolecular forces, and reactions can be explained in terms of potential energy due to electromagnetic forces and kinetic energy that corresponds to temperature. In this model, the electrostatic attractive forces between oppositely charged particles and the electrostatic repulsive forces between identical charges govern the energies of electrons in atoms, in bonds, and in condensed phases.

Nuclear Structure and Reactions

At the nuclear level, however, in addition to electromagnetic forces, other forces of nature are in effect. In any nucleus with more than one proton, the positive charge repulsion is overcome by another attractive force called the *strong force,* which is only effective at extremely small distances (within the nucleus). This force is necessary for elements with more than one proton to exist! In this activity we will use the term *nucleon* to refer to **both** protons and neutrons in the nuclei of isotopes. One other force that must be considered when studying nuclear reactions is that which is responsible for *beta decay* that results in the conversion of a neutron to a proton and an electron. This type of radioactivity, and other nuclear reactions, will be discussed in a later activity.

In nuclear reactions, these additional forces account for enormous energy changes that will be examined in this activity. Remember from your initial introduction that the term *isotope* refers to an atom with a specific number of protons, electrons, *and* neutrons. The symbol for an isotope is written $^{A}_{Z}\text{X}^{+/-m}$, where **X** is the element's atomic symbol; *A* is its *mass number,* equal to the sum of its nucleons (protons and neutrons); *Z* is its *atomic number,* equal to the total number of protons; and $+/- m$ indicates the net charge of the ion, determined by the difference in the number of protons and electrons.

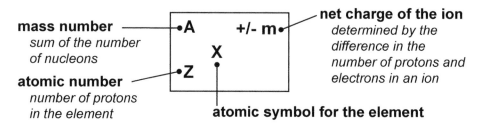

A *nuclide,* which is comprised of nucleons, is similar to an isotope, but differs in that it describes **the atomic nucleus** with a specific number of protons and neutrons, but **no electrons.** Thus the nuclide $^{12}_{6}C$ can be interpreted as carbon (C) with 12 nucleons, 6 protons (and therefore 6 neutrons), with no net charge. The *binding energy of a nuclide* is the energy required to separate it into its component protons and neutrons. Binding energies of nuclides determine the energy released in nuclear reactions and provide insights regarding the relative stability of elements and their isotopes.

Nuclear Binding Energy

Since the advent of Albert Einstein's Theory of Special Relativity, mass and energy are considered to be different manifestations of the same quantity. The equation $E = mc^2$ (where c is the speed of light) provides the conversion between mass, *m,* and energy, *E.* In most situations, e.g., chemical reactions, energy changes are so small that corresponding changes in mass cannot be observed. The binding energies of nuclides are significantly larger, so it is useful to view binding energies as changes in mass. Consequently, the binding energy of a nuclide, E_b, and the binding energy per nucleon, $(E_b)/A$, can be calculated by the following procedure, where A is the mass number. Step 5 in this procedure is very important; in order to see how strongly *one* nucleon (proton or neutron) is held within a nuclide, it is necessary to divide the total binding energy of the nuclide by *the number of nucleons* that comprise it.

(1) $E_b = E_{separated\ nucleons} - E_{nuclide}$

(2) $E_b = (m_{separated\ nucleons})(c^2) - (m_{nuclide})(c^2)$

(3) $E_b = (m_{separated\ nucleons} - m_{nuclide})(c^2)$

(4) $E_b = \Delta mc^2$ (*This is the binding energy per* **nuclide.**)

(5) $\dfrac{E_b}{A}$ (*This is the binding energy per* **nucleon.**)

| MODEL 1 | CALCULATION OF THE BINDING ENERGY PER NUCLEON |

The paradox of isotopic mass

particle	mass	
proton (p$^+$)	1.007276 amu	or 1.007276 g/mol
neutron (n^0)	1.008665 amu	or 1.008665 g/mol
electron (e$^-$)	0.00054858 amu	or 0.00054858 g/mol

Find the masses of the particles in carbon-12:

particle	number of particles	×	mass (g/mol)	=	total particle mass (g/mol)
p$^+$	6	×	1.007276	=	6.043656
n^0	6	×	1.008665	=	6.05199
e$^-$	6	×	0.00054858	=	0.00329148

Add them:

6.043656

6.05199

+   0.00329148
_____

total mass   **12.09893748 g/mol** → But, by **definition**, the isotope mass of carbon-12 is 12.000000 amu, or 12.000000 g/mol!

Calculate the binding energy per nucleon of the carbon-12 isotope:

Use the procedure in the information section:

(3) $E_b = (m_{separated\ nucleons} - m_{nuclide})(c^2)$

(4) $E_b = \Delta mc^2$

(5) $E_{b/A} = \dfrac{E_b}{A}$

Calculate the mass of the separated nucleons:

$m_{separated\ nucleons} =$ *calculated total particle mass – electron mass*

$m_{separated\ nucleons} =$ (12.09893748 g/mol) – (6)(0.00054858 g/mol) = **12.095646 g/mol**

and the mass of the nuclide:

$m_{nuclide} =$ *isotope mass as given – electron mass*

$m_{nuclide} =$ (12.000000 g/mol) – (6)(0.00054858 g/mol) = **11.996709 g/mol**

This difference in mass (Δm) is also known as the *mass defect.*

Calculate the difference in mass in order to find the binding energy:

$\Delta m = m_{separated\ nucleons} - m_{nuclide}$

$\Delta m =$ (12.09893748 g/mol) – (11.996709 g/mol) = 0.098937 g/mol or
9.893748 × 10⁻⁵ kg/mol

Now we can calculate the binding energy of the isotope: $E_b = \Delta mc^2$

$E_b = \Delta m \times c^2$

$E_b =$ (9.8937 × 10⁻⁵ kg/mol) × (2.99792 × 10⁸ m/sec)² = **8.892030 × 10¹² J/mol**

Finally, divide by the mass number to find the binding energy **per nucleon**: $E_{b/A} = \dfrac{E_b}{A}$

$E_{b/A} =$ (8.892030 × 10¹² J/mol)(1 kJ/1000 J)/(12 nucleons) = **7.410025 × 10⁸ kJ/mol**

1. In calculating the mass of the nuclide, why was it necessary to subtract the mass of six electrons?

2. In calculating the mass of any nuclide from data about isotopic masses, what must be done before proceeding to the calculation of the mass defect?

3. When calculating binding energy in J/mol, what must be done to make the units work out correctly?

4. How does the binding energy for the carbon-12 nucleus compare to energy changes observed in chemical reactions involving carbon-12?

5. What other differences can you see between chemical and nuclear reactions?

EXERCISE

1. Calculate the binding energy per nucleon for iron-56 and iron-58. Place these points, along with the values you calculated, on the graph in **Model 2**. Identify the isotope that is more stable.

   For this calculation you need precise values for the mass of a proton, a neutron, and the iron isotopes. Remember that $1 \text{ J} = 1 \text{ kg m}^2/\text{s}^2$, and note that the units on the graph are 10^8 kJ/mol.

proton mass = 1.007276 g/mol
neutron mass = 1.008665 g/mol
electron mass = 5.4858×10^{-4} g/mol

Fe-56 isotopic mass = 55.934994 g/mol
Z = 26, N = 30, A = 56

Fe-58 isotopic mass = 57.933275 g/mol
Z = 26, N = 32, A = 58

Binding energy per nucleon:

iron-56

iron-58

Which isotope is more stable?

MODEL 2 COMPARING BINDING ENERGIES OF NUCLEONS

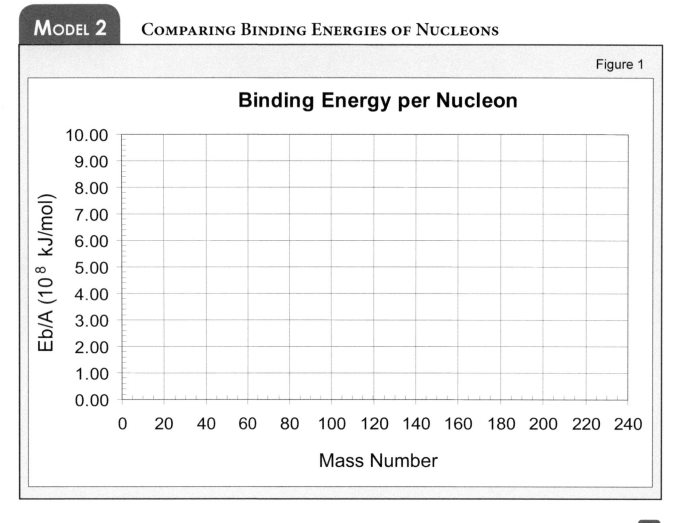

Figure 1

Binding Energy per Nucleon

KEY QUESTIONS

6. Of all the isotopes in **Model 2**, which has the smallest binding energy per nucleon?

7. Why do you think the binding energy per nucleon is small when the mass number is very small?

8. Using the ideas of a repulsive electromagnetic force and an attractive nuclear force, why do you think the binding energy per nucleon first increases and then decreases as the number of nucleons increases past some intermediate value?

9. Which nuclides are more stable: those with a large binding energy per nucleon or those with a small binding energy per nucleon? Explain.

10. Is the reaction that converts a less stable nuclide to a more stable nuclide exothermic or endothermic? Would you expect the free energy change for this reaction to be positive or negative? Explain.

11. Taking into account your answer to Key Question 10, which of those isotopes in the model would you expect to be most abundant at equilibrium? Explain.

12. Why is the distribution of isotopes not the equilibrium distribution, e.g., why do we still have hydrogen, helium, and uranium?

13. Will a nuclide with a small mass number become more stable by combining to form a larger nuclide or by splitting into smaller nuclides? Explain.

14. Will a nuclide with a large mass number become more stable by combining with another nuclide or by splitting into smaller nuclides? Explain.

15. There are two kinds of atomic bombs: hydrogen and uranium. A hydrogen bomb is much more destructive. Why do you think more energy is released from a hydrogen bomb than a uranium bomb?

INFORMATION

All elements are considered thermodynamically unstable with respect to iron because the free energy change required to produce iron from these elements is negative. Therefore, while all elements *should* spontaneously convert to iron, they do not do so when the **activation energy** for such a spontaneous conversion is large.

Light nuclides can come together to form more stable heavier nuclides. This process is called *fusion*, and it is the source of energy released by the sun. Heavy nuclides can split to form more stable lighter nuclides. This process is called *fission*, and it is the source of energy in nuclear power generation stations.

EXERCISES

2. Calculate the energy released when one mole of ^2H is converted into ^3He. The nuclear reaction equation is $2\,^2H \rightarrow\,^3He$ + neutron. The nuclide masses are $^2H = 2.0141$ g/mol, $^3He = 3.0160$ g/mol.

3. How much energy is released per mole of uranium in the following nuclear reaction? The nuclide masses in g/mol are U (235.0439), Mo (99.9076), and Sn (133.9125).

$$^{235}U + n \longrightarrow {}^{100}Mo + {}^{134}Sn + 2n$$

WHAT DO YOU THINK NOW?

Is it possible to turn lead into gold? If no, explain why not. If yes, explain how it might be done. In your explanation, mention whether gold is more or less stable than lead.

WHY?

Not all elements and isotopes are stable. Radioactivity is the spontaneous emission of particles or energy from the nucleus of an unstable isotope. Such isotopes are called *radioisotopes*. These spontaneous nuclear reactions produce **more stable** elements and isotopes from those that are *less stable*. Nuclear chemistry is the subdiscipline of chemistry that involves the study of these reactions and their applications to research, technology, and medicine. An understanding of nuclear chemistry is essential when discussing issues associated with radioactivity, nuclear power generation, disposal of radioactive waste, and the intelligent use of radiation in medicine.

WHAT DO YOU THINK?

Are nuclear reactions similar to chemical reactions? Identify any similarities and differences.

LEARNING OBJECTIVE

- Understand what happens during the radioactive decay of an isotope

SUCCESS CRITERION

- Write nuclear reaction equations to describe how nuclei change through fission, fusion, and the emission or capture of particles

PREREQUISITES

02-1   *Atoms, Isotopes, and Ions*

22-1   *Nuclear Chemistry: Binding Energy*

TASKS

1. Examine the model on the following page and, for each example, identify the quantities that are conserved in the nuclear reaction. The *Key Questions* will help you with this identification.

2. Using the conservation rules that you have identified, complete the second example that is given in each row of the model.

The symbol for a nuclide, like an atomic symbol, includes the symbol for the element, a superscript for the mass number, A, and a subscript for the atomic number, Z. (Remember that the mass number is equal to the total number of protons and neutrons contained in the nucleus, and that the atomic number specifies the number of protons the nuclide contains, which defines its charge.) Since superscripts and subscripts are also used in symbols for subatomic particles like electrons and positrons, the subscript Z now is called the *charge number*.

MODEL — SPONTANEOUS NUCLEAR REACTIONS

Process	Symbol	ΔZ	ΔA	Examples
alpha particle emission	α or ^4_2He	−2	−4	$^{238}_{92}U \rightarrow {}^{234}_{90}Th + \alpha$ $\square \rightarrow {}^{222}_{86}Rn + \alpha$
beta particle emission	β or $^0_{-1}e$	+1	0	$^{131}_{53}I \rightarrow {}^{131}_{54}Xe + \beta$ $\square \rightarrow {}^{32}_{16}S + \beta$
positron emission	β^+ or $^0_{+1}e$	−1	0	$^{22}_{11}Na \rightarrow {}^{22}_{10}Ne + \beta^+$ $\square \rightarrow {}^{15}_{7}N + \beta^+$
neutron emission	^1_0n	0	−1	$^{12}_{4}Be \rightarrow {}^{11}_{4}Be + {}^1_0n$ $\square \rightarrow {}^{4}_{2}He + {}^1_0n$
gamma ray emission, excited state decays	γ	0	0	$^{57}_{26}Fe^* \rightarrow {}^{57}_{26}Fe + \gamma$ $\square \rightarrow {}^{119}_{50}Sn + \gamma$
electron capture (K-capture)	$^0_{-1}e$	−1	0	$^{7}_{4}Be + {}^0_{-1}e \rightarrow {}^{7}_{3}Li$ $\square + {}^0_{-1}e \rightarrow {}^{40}_{18}Ar$

Process	Symbol	ΔZ	ΔA	Examples
fission		various	various	$^{236}_{92}U \rightarrow {}^{141}_{56}Ba + {}^{92}_{36}Kr + 3\,{}^{1}_{0}n$ $\boxed{} \rightarrow {}^{103}_{42}Mo + {}^{131}_{50}Sn + 2\,{}^{1}_{0}n$
fusion		various	various	${}^{2}_{1}H + {}^{2}_{1}H \rightarrow {}^{3}_{2}He + {}^{1}_{0}n$ $\boxed{} + {}^{3}_{2}He \rightarrow {}^{4}_{2}He + {}^{0}_{+1}e$

KEY QUESTIONS

1. What information is given by the superscripts and subscripts on the symbols used in the model?

2. What is the change in the charge number when an alpha particle is emitted from a nuclide? What is the change in its mass number?

3. What is the change in the charge number when a beta particle is emitted from a nuclide? What is the change in its mass number?

4. What is the relationship between the mass numbers of the reactants and the mass numbers of the products in a nuclear reaction?

5. What is the relationship between the charge numbers of the reactants and the charge numbers of the products in a nuclear reaction?

6. Can a nuclear reaction appear to turn a proton into a neutron, or a neutron into a proton? If it cannot, explain why not. If it can, describe how.

7. Using your knowledge of chemical reactions, identify what factor in a nuclear reaction determines whether it is spontaneous or not.

8. Using your knowledge of chemical reactions, identify what determines whether a nuclear reaction has a small or large rate constant.

EXERCISES

1. Write an equation describing the radioactive decay of each of the following isotopes. The type of decay is given in parentheses.

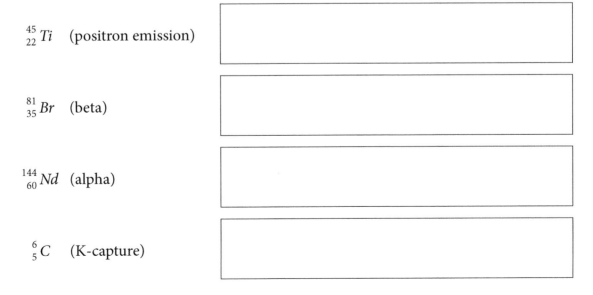

$^{45}_{22}Ti$ (positron emission)

$^{81}_{35}Br$ (beta)

$^{144}_{60}Nd$ (alpha)

$^{6}_{5}C$ (K-capture)

Wʜᴀᴛ Dᴏ Yᴏᴜ Tʜɪɴᴋ Nᴏᴡ?

Identify the similarities and differences between chemical reactions and nuclear reactions.

ACTIVITY 22-3 Rates of Radioactive Decay

WHAT DO YOU THINK?

Radioactive iodine-131 is used to treat thyroid cancer. A patient is given 20.0 mg of iodine-131 in the form of NaI, and after 8 days only 10 mg remains. How long, from the time of the initial dose, do you think it will it take until only 5 mg remain?

    a. 12 days

    b. 16 days

    c. 20 days

    d. cannot be determined from the information given

LEARNING OBJECTIVES

- Understand how levels of radioactivity decrease with time
- Understand how radioactivity can be used to determine the age of materials

SUCCESS CRITERIA

- Relate the amount of radioactive material remaining after some period of time to the half-life of the radioactive isotope and the rate constant for its decay
- Estimate the age of materials from their radioactivity

PREREQUISITES

13-1   *Rates of Chemical Reactions*

22-2   *Radioactivity*

TASKS

1. Examine the model and complete Table 1 by entering *N*, the number of unstable nuclides remaining at each point in time, and ln(*N*) in the last two columns, respectively.

MODEL 1 TIME EVOLUTION OF NUCLEAR DECAY

In the chart below, ten unstable nuclides are represented by white circles. They decay spontaneously by some mechanism to produce stable nuclides, which are indicated by dark circles.

Table 1

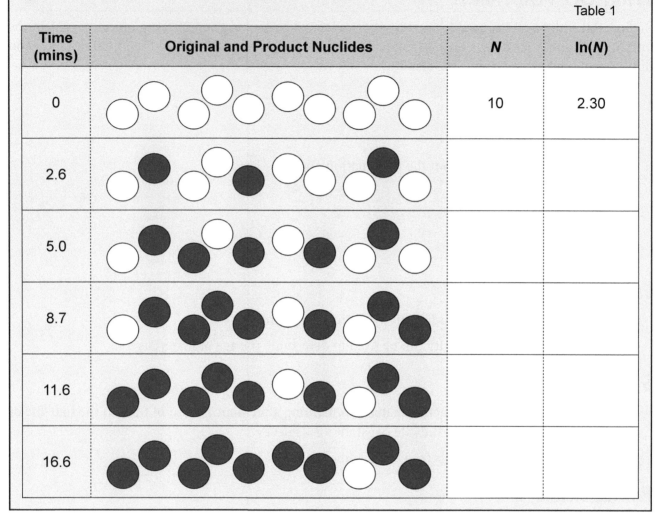

Time (mins)	Original and Product Nuclides	N	ln(N)
0		10	2.30
2.6			
5.0			
8.7			
11.6			
16.6			

2. Make two graphs of the data (see the next page): Graph 1 with N plotted on the y-axis and time plotted on the x-axis, and Graph 2 with $\ln(N)$ plotted on the y-axis and time plotted on the x-axis. (Remember that high quality graphs have titles, labels on the axes, data points shown with small circles around them, and a smooth line drawn through the data points.)

3. On your graphs, mark the points on the x-axis where the fraction of unstable nuclides remaining is equal to $\frac{1}{2}$, $\frac{1}{4}$, and $\frac{1}{8}$ of the initial amount.

Graph 1

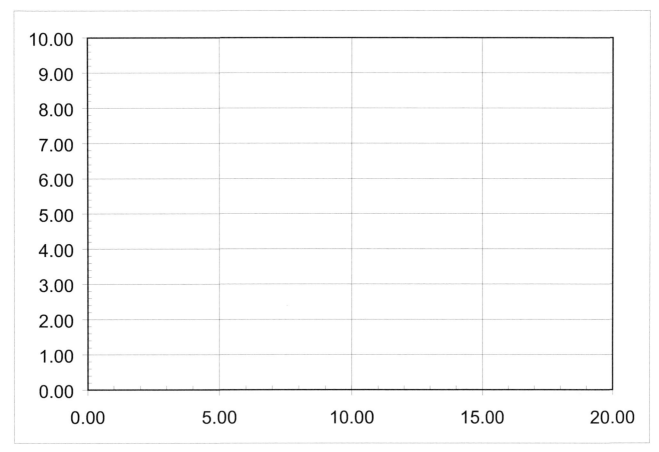

Graph 2

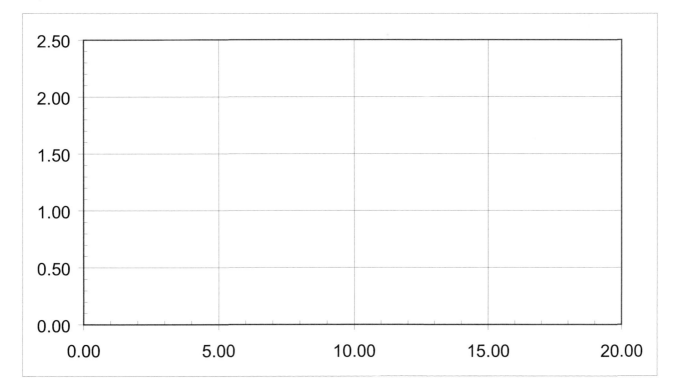

KEY QUESTIONS

1. According to Graph 1, how long does it take for half of the radioactive nuclides in the model to decay? This time is referred to as the nuclide's *half-life*.

2. Does the half-life of the nuclide depend on when the first measurement is taken? Explain how the information in Graph 1 supports your answer.

3. How can the data in the model be used to identify whether radioactive decay is zero order, first order, or second order?

4. What is the mathematical equation that represents the straight line that can be drawn through the data points in Graph 2?

5. How can the rate constant for this radioactive decay be obtained from Graph 2?

Use the following information for Exercises 1–4: Two radioactive isotopes, A and B, have decay rate constants k_A and k_B, respectively, where k_A is larger than k_B. N_{A_0} and N_{B_0} are the number of nuclides present at $t = 0$ for each of these isotopes. The decay rate R is the number of decay events per second, where

$$R = \Delta N / \Delta t = -k\,N$$

1. Sketch one graph that contains both a curve for $N_A(t)$ and a curve for $N_B(t)$. Compare the number of radioactive nuclides present as a function of time for both of these isotopes; start at t = 0, and let the initial number of A isotopes and B isotopes be equal. (It might be helpful to look at the graphs you constructed based on the information in the model, and think about how they might be different if the decay rate were faster.)

2. Sketch one graph with a curve that contains $\ln[N_A(t)]$ and a curve for $\ln[N_B(t)]$, as a function of time. Again, start at $t = 0$, with same number of A isotopes and B isotopes.

3. Identify the isotope with the longer half-life.

4. Identify the isotope with the faster decay rate when the amounts of A and B present are equal.

5. Using your answer to Key Question 4, derive the following relationship between the half-life of a nuclide and the rate constant: $t_{1/2} = \ln(2) / k$.

6. Derive your answer to Key Question 4 again, using this equation, which describes the data in Graph 1: $N = N_o e^{-kt}$

7. Pd-100 has a half-life of 3.6 days. Using your answer to Exercise 5, calculate a value for the decay rate constant, k, in units of s^{-1}.

8. Determine the half-life of a radioactive nuclide if, after 3.75 hrs, only 1/16 of the initial amount remains unchanged.

9. Estimate how many half-lives must pass before less than 1% of a radioisotope remains.

WHAT DO YOU THINK NOW?

1. A patient is given 20.0 mg of iodine-131 in the form of NaI. After 8 days, only 10 mg remains. How long, from the time of the initial dose, do you think it will it take until only 5 mg remain?

    a.   12 days

    b.   16 days

    c.   20 days

    d.   cannot be determined from the information given

2. How long will it take until only 0.1% of the iodine-131 remains?

PROBLEMS

1. A 0.145 pg sample of cobalt-56 produces 162 β-particles per second. What is the half-life of this isotope?

2. A wooden tool is found at an archaeological site. Estimate the age of the tool using the following information: A 100 gram sample of the wood emits 1025 β-particles per minute, due to the decay of carbon-14. The decay rate of carbon-14 in living trees is 15.3 per minute per gram. Carbon-14 has a half-life of 5,730 years.

3. Potassium-40 decays into argon-40 by electron capture, with a half-life of 1.28×10^9 years. This decay forms the basis of the potassium-argon dating method that is used with igneous rocks. This method assumes that no argon was trapped in the rock when it was formed from the molten material.

A meteorite was found to have equal amounts of ^{40}K and ^{40}Ar.

a. Of the potassium-40 that was originally present, what fraction has decayed?

b. How long ago was this meteorite formed?

c. Another meteorite had a ^{40}K to ^{40}Ar ratio of 0.33. How long ago was this meteorite formed?

This page intentionally left blank.

INTERNATIONAL SYSTEM OF UNITS (SI)

http://physics.nist.gov/cuu/Units

SI PREFIXES

Factor	Name	Symbol
10^{24}	yotta	Y
10^{21}	zetta	Z
10^{18}	exa	E
10^{15}	peta	P
10^{12}	tera	T
10^{9}	giga	G
10^{6}	mega	M

Factor	Name	Symbol
10^{3}	kilo	k
10^{2}	hecto	h
10^{1}	deka	da
10^{-1}	deci	d
10^{-2}	centi	c
10^{-3}	milli	m
10^{-6}	micro	μ

Factor	Name	Symbol
10^{-9}	nano	n
10^{-12}	pico	p
10^{-15}	femto	f
10^{-18}	atto	a
10^{-21}	zepto	z
10^{-24}	yocto	y

DEFINITIONS OF SI BASE UNITS

Base Quantity	Name	Symbol	Definition
length	meter	m	The meter is the length of the path travelled by light in vacuum during a time interval of 1/299 792 458 of a second.
mass	kilogram	kg	The kilogram is the unit of mass; it is equal to the mass of the international prototype of the kilogram.
time	second	s	The second is the duration of 9 192 631 770 periods of the radiation corresponding to the transition between the two hyperfine levels of the ground state of the cesium 133 atom.
electric current	ampere	A	The ampere is that constant current which, if maintained in two straight parallel conductors of infinite length, of negligible circular cross-section, and placed 1 meter apart in vacuum, would produce between these conductors a force equal to 2×10^{-7} newton per meter of length.
thermodynamic temperature	kelvin	K	The kelvin, unit of thermodynamic temperature, is the fraction 1/273.16 of the thermodynamic temperature of the triple point of water.
amount of substance	mole	mol	1. The mole is the amount of substance of a system which contains as many elementary entities as there are atoms in 0.012 kilogram of carbon 12; its symbol is "mol." 2. When the mole is used, the elementary entities must be specified and may be atoms, molecules, ions, electrons, other particles, or specified groups of such particles.
luminous intensity	candela	cd	The candela is the luminous intensity, in a given direction, of a source that emits monochromatic radiation of frequency 540×10^{12} hertz and that has a radiant intensity in that direction of 1/683 watt per steradian.

Units Derived from SI Base Units

Derived quantity	Name	Symbol	Expression in terms of other SI units	Expression in terms of SI base units
area	square meter	m^2		$m \cdot m$
volume	cubic meter	m^3		$m \cdot m \cdot m$
plane angle	radian	rad		$m\, m^{-1}$
solid angle	steradian	sr		$m^2\, m^{-2}$
speed	meter per second	m/s		m/s
acceleration	meter per second squared	m/s^2		m/s^2
frequency	hertz	Hz		s^{-1}
force	newton	N		$kg \cdot m\, s^{-2}$
pressure	pascal	Pa	N/m^2	$kg\, m^{-1} \cdot s^{-2}$
energy, work	joule	J	N·m	$kg\, m^2 \cdot s^{-2}$
power	watt	W	J/s	$kg\, m^2 \cdot s^{-3}$
electrical charge	coulomb	C		A s
electrical potential, electromotive force	volt	V	W/A	$m^2 \cdot kg \cdot s^{-3} \cdot A^{-1}$
electric field strength	volt per meter	V/m	V/m	$m \cdot kg \cdot s^{-3} \cdot A^{-1}$
capacitance	farad	F	C/V	$m^{-2} \cdot kg^{-1} \cdot s^4 \cdot A^2$
electrical resistance	ohm	Ω	V/A	$m^2 \cdot kg \cdot s^{-3} \cdot A^{-2}$
activity of a radionuclide	becquerel	Bq		s^{-1}
absorbed radiation dose	gray	Gy	J/kg	$m^2 \cdot s^{-2}$
biological equivalent dose	sievert	Sv	J/kg	$m^2 \cdot s^{-2}$

Other Units, Abbreviations, and Equivalence Statements

LENGTH		
Name	**Symbol**	**Equivalence Statement**
ångstrom	Å	$Å = 0.1\ nm = 10^{-10}\ m$
inch	in	1 in = 2.54 cm
foot	ft	1 ft = 12 in
mile	mi	1 mi = 1.609344 km; 1 mi = 5280 ft
MASS		
Name	**Symbol**	**Equivalence Statement**
atomic mass unit	u or amu	$1\ u = 1.6605388 \times 10^{-27}\ kg$
pound	lb	1 lb = 0.45359237 kg

Mass (con't)

metric ton or tonne	t	1 t = 1000 kg
ton	T	1 T = 2000 lb

Volume

Name	Symbol	Equivalence Statement
liter	L	$1 \text{ L} = 10^{-3} \text{ m}^3$
gallon	gal	1 gal = 3.78541 L
quart	qt	4 qt = 1 gal
pint	pt	2 pt = 1 qt
ounce	oz	16 oz = 1 pt
tablespoon	T or tbsp	2 tbsp = 1 oz
teaspoon	t or tsp	3 tsp = 1 tbsp

Pressure

Name	Symbol	Equivalence Statement
bar	bar	$1 \text{ bar} = 10^5 \text{ Pa}$
atmosphere	atm	1 atm = 101.325 kPa; 1 atm = 1.01325 bar
torr	torr	760 torr = 1 atm
pounds per square inch	psi or lb/in^2	14.696 psi = 1 atm

Energy

Name	Symbol	Equivalence Statement
electron volt	eV	$1 \text{ eV} = 1.60218 \times 10^{-19} \text{ J}$
calorie	cal	1 cal = 4.184 J
nutritional calorie	Cal	1 Cal = 1000 cal = 1 kcal

Temperature

Name	Symbol	Equivalence Statement
Celsius temperature	°C	$T_{°C} = T_K (1 \text{ °C} / 1 \text{ K}) - 273.16 \text{ °C}$
Fahrenheit temperature	°F	$T_{°F} = T_{°C} (180 \text{ °F} / 100 \text{ °C}) + 32 \text{ °F}$

Time

Name	Symbol	Equivalence Statement
minute	min	1 min = 60 s
hour	h	1 h = 60 min
day	d	1 d = 24 h

ANGLE		
Name	**Symbol**	**Equivalence Statement**
degree	°	$360° = 2\pi$ rad
minute	′	$60′ = 1°$
second	″	$60″ = 1′$

PHYSICAL CONSTANTS

http://nist.gov/cuu/Constants/index.html

Quantity	Value*
speed of light in a vacuum (c)	2.99792458×10^8 m/s
elementary charge (+e for proton, –e for electron)	$1.6021765 \times 10^{-19}$ C
electron rest mass (m_e)	$9.1093822 \times 10^{-31}$ kg
proton rest mass (m_p)	$1.67262164 \times 10^{-27}$ kg
neutron rest mass (m_n)	$1.67492721 \times 10^{-27}$ kg
Planck constant (h)	$6.6260690 \times 10^{-34}$ Js
Avogadro constant (N_A or L)	6.0221418×10^{23} /mol
Faraday constant (**F**)	96,485.340 C/mol
gas constant (R)	8.31447 J K^{-1} mol^{-1} 0.082057 L atm mol^{-1} K^{-1}
Boltzman constant (k)	1.380650×10^{-23} J/K

Values are rounded to include only one uncertain digit.

STANDARD REDUCTION POTENTIALS ARE FOUND ON PAGE 408 (ACTIVITY 18-1).

Secrets for Success in General Chemistry

Students who completed General Chemistry with grades between A- and A+ were interviewed to find out *how* they did it. Here's what they had to say...

IN GENERAL

- Plan to devote 9-12 hours each week to studying chemistry, make a schedule of study times, otherwise you will never put in this much time each week. It just isn't that much fun!
- Be sure to get enough sleep, exercise, and good nutrition always, but particularly before exams
- Understand the material from each class before the next class
- If you get stuck, use your textbook as a resource, get help from your friends and classmates, or go to your instructor
- Study for exams by identifying the key ideas, understanding what they mean, and seeing how they are used in solving problems. Review all the homework problems and connect them to the key ideas.
- Do not simply read and reread the textbook or class notes, rather use these resources to answer specific questions that develop as you do the assigned homework

BEFORE CLASS

- Spend a few minutes looking over the reading assignment to identify key points because you learn by building on what you already know

DURING CLASS

- Listen carefully, take notes, and mark issues that are not clear
- Work with others to complete and understand the *Foundations of Chemistry* activity and in-class problems

AFTER CLASS

- Review your notes, use your textbook (or the instructor's notes if available) to complete them and clarify issues
- Make sure you understand the in-class activity and sample problems
- Carefully complete the reading assignment. Take notes to highlight the important points.
- Test your understanding by completing the homework assignment. Use your textbook as a resource to resolve any issues that arise. Anything that you are asked to do is explained in class or in the reading assignment. You just need to understand and put the ideas together.

AFTER YOU HAVE COMPLETED AN ASSIGNMENT

- If your instructor does not score your homework, check your answers and solutions against answers or solutions that are posted, and grade yourself
- Do not use posted solutions as examples that you read and attempt to memorize. The understanding that you need for success on exams is developed by figuring things out before the solutions are posted, not by reading and memorizing the solutions that someone else prepared.
- If you don't understand a solution, analyze similar examples in your textbook, get help from friends and classmates, or see your instructor

APPENDIX C
How Your Brain Works: Implications for Learning Chemistry Efficiently

Your brain has a short-term memory or working memory that processes incoming and outgoing information. The capacity of this working memory is small; it can only deal with five to nine pieces of information. The working memory can be expanded by using paper to take notes and work out solutions to problems. With ideas and notes on paper, many items and the connections between them can be visualized.[1]

Your brain has a perception filter that is controlled by prior knowledge and experiences that are stored in long-term memory.[1] The perception filter restricts the sensory information that reaches the working memory. The information that does reach working memory subsequently is stored in long-term memory if properly reinforced by repetition. Since learning depends on what you already know, you need to go over material repetitively, learning a little bit more each time. If you don't understand something, come back to it the next day because your brain continues to work and make connections while you sleep. Do not try to cover too much material in too short of a time because the rate at which you learn is limited by the perception filter, the capacity of working memory, and the rate your brain processes new information. It needs time to consolidate and connect ideas, and a good night's sleep helps. Spread the time you devote to learning chemistry over the entire week; don't concentrate it in a few days.

Your brain has a librarian function. There is no way that words or actions of a teacher can be stored directly in long-term memory and automatically be understood and utilized by the learner. There must be a conversion of this external language into your knowledge system with added connections so the knowledge can be retrieved and used when needed.[2,3] It is the librarian that executes this task. Initially the librarian stores information as disconnected pieces. To make use of this knowledge, the facts, vocabulary, equations, procedures, concepts, and contexts need to be interconnected in hierarchical structures. The following four actions on your part are most important to assist the librarian in making these connections and building these structures.[4]

- Identify the important concepts covered in an assignment and the connections between those concepts.

- Think about a concept in multiple representations: as a verbal statement, using symbols or an equation, and as a picture, diagram, or graph.

- Reflect on the concepts and procedures used in textbook examples or in solving a homework problem or answering a question: what concepts were needed and why were they needed, what were the steps in the procedures and why were those steps necessary, and how are the concepts related to the procedures?

- Solve many different problems with different contexts because repetition is essential. Analyze these different situations to identify similarities and differences in order to build a library of contexts to draw upon by analogy.

Your brain needs a reflector/analyzer. The only way to improve performance is to think about what you have done well, what needs to be improved, what has been learned, and what is not yet understood. Such reflection and analysis is essential for continual improvement, growth, and success.[5]

---

1 Johnstone, A. H. *J. Chem. Ed.* 1997, *74*, 262-268.

2 Newell, A.; Simon, H. A. *Human Problem Solving*; Prentice-Hall: Englewood Cliffs, NJ, 1972.

3 Simon, H. A. In *Problem Solving and Education: Issues in Teaching and Research*; Tuma, D. T., Reif, F., Eds.; Lawrence Erlbaum Associates: Hillsdale, NJ, 1980, p 81-96.

4 Hanson, D. M. In *Process-Oriented Guided-Inquiry Learning*; Moog, R. S., Spencer, J. N., Eds.; American Chemical Society: Washington, DC, 2008, p 14-25.

5 *How People Learn: Brain, Mind, Experience, and School*; Bransford, J. D.; Brown, A. L.; Cocking, R. R., Eds.; National Academy Press: Washington, D,C, 2000.

Index

Symbols

α-amino acids 450
α-helix 455
β-sheets 455

A

accuracy 12
acids 89, 91
 carboxylic acids 90
 strong acids 89
 weak acids 89
activated complex. *See* transition state
activation energy 312, 466
active agents 356
addition polymerization 443
addition polymers 443
amide linkage 450
amino acids 450
amu 18
analytes 105
angular momentum quantum number (ℓ) 150
angular nodes 150
anode 406
aqueous solution 73, 76
Arrhenius Equation 316
atom 17
atomic mass unit 23
atomic orbitals 149
Atomic Symbol Notation 18, 19
Aufbau Principle 162
average rate of a reaction 293
Avogadro's number 23
azimuthal quantum number (ℓ) *See* angular momentum quantum number (ℓ)

B

balance chemical reaction equations 58
base 89
 Brønsted-Lowry base 89
 Strong bases 90
 Weak bases 90
beta decay 458
binary covalent compounds 32
binary ionic compounds 33
binding energy 457
binding energy of a nuclide 458
Boltzmann's constant 327
bond energy 128
bond order 128

Born-Haber cycle 134
boundary surface 149
Brønsted-Lowry 89
Brønsted-Lowry definition 363
buffers 379

C

catalysts 312
cathode 406
cation 18
charge number 469
chelating agent 423
chemical reaction mechanism 303
Clausius-Clapeyron equation 272
colligative properties 287
compound 17, 28
 binary 32
 covalent 32
 ionic 17
conjugate acid of the base 92
conjugate base of the acid 92
conversion factor 4
conversion factors 4
cooling curve 266
coordinate covalent bond 423
coordination complex. *See* coordination compounds
coordination compounds 422
coordination number 422, 423
corrosion 94
coulomb 408
counter ions 423
covalent bonding 258
covalent bonding interactions 258
crosslinking 445
crystal field theory (CFT) 428

D

differential rate law. *See* rate law
dimensional analysis. *See also* units: analysis
dipole-dipole interactions 259
dissociation 84
dissociation energy 261
Dmitri Mendeleev 28
double replacement reactions 85

E

electrochemical cells 406
electrode 406
electrolytic cell 416

List of Models